Lecture Notes in Mathematics

Edited by A. Dold, B. Eckmann and F. Takens

1419

P.S. Bullen P.Y. Lee J.L. Mawhin
P. Muldowney W.F. Pfeffer (Eds.)

New Integrals

Proceedings of the Henstock Conference
held in Coleraine, Northern Ireland, August 9–12, 1988

Springer-Verlag
Berlin Heidelberg New York London Paris Tokyo Hong Kong

Editors

Peter S. Bullen
Department of Mathematics, University of British Columbia
Vancouver, British Columbia
Canada V6T 1Y4

Peng Yee Lee
Department of Mathematics, National University of Singapore
Singapore 0511, Republic of Singapore

Jean L. Mawhin
Département de Mathématiques, Université Catholique de Louvain
2 chemin du Cyclotron, 1348 Louvain-La-Neuve, Belgium

Patrick Muldowney
Magee College, University of Ulster
Londonderry BT48 7JL, Northern Ireland

Washek F. Pfeffer
Department of Mathematics, University of California
Davis, CA 95616, USA

Mathematics Subject Classification (1980): MR26A39

ISBN 3-540-52322-7 Springer-Verlag Berlin Heidelberg New York
ISBN 0-387-52322-7 Springer-Verlag New York Berlin Heidelberg

Printed in Germany

Printing and binding: Druckhaus Beltz, Hemsbach/Bergstr.
2146/3140-543210 – Printed on acid-free paper

Lecture Notes in Mathematics

For information about Vols. 1–1201 please contact your bookseller or Springer-Verlag

Vol. 1202: A. Dür, Möbius Functions, Incidence Algebras and Power Series Representations. XI, 134 pages. 1986.

Vol. 1203: Stochastic Processes and Their Applications. Proceedings, 1985. Edited by K. Itô and T. Hida. VI, 222 pages. 1986.

Vol. 1204: Séminaire de Probabilités XX, 1984/85. Proceedings. Edité par J. Azéma et M. Yor. V, 639 pages. 1986.

Vol. 1205: B.Z. Moroz, Analytic Arithmetic in Algebraic Number Fields. VII, 177 pages. 1986.

Vol. 1206: Probability and Analysis, Varenna (Como) 1985. Seminar. Edited by G. Letta and M. Pratelli. VIII, 280 pages. 1986.

Vol. 1207: P.H. Bérard, Spectral Geometry: Direct and Inverse Problems. With an Appendix by G. Besson. XIII, 272 pages. 1986.

Vol. 1208: S. Kaijser, J.W. Pelletier, Interpolation Functors and Duality. IV, 167 pages. 1986.

Vol. 1209: Differential Geometry, Peñíscola 1985. Proceedings. Edited by A.M. Naveira, A. Ferrández and F. Mascaró. VIII, 306 pages. 1986.

Vol. 1210: Probability Measures on Groups VIII. Proceedings, 1985. Edited by H. Heyer. X, 386 pages. 1986.

Vol. 1211: M.B. Sevryuk, Reversible Systems. V, 319 pages. 1986.

Vol. 1212: Stochastic Spatial Processes. Proceedings, 1984. Edited by P. Tautu. VIII, 311 pages. 1986.

Vol. 1213: L.G. Lewis, Jr., J.P. May, M. Steinberger, Equivariant Stable Homotopy Theory. IX, 538 pages. 1986.

Vol. 1214: Global Analysis – Studies and Applications II. Edited by Yu.G. Borisovich and Yu.E. Gliklikh. V, 275 pages. 1986.

Vol. 1215: Lectures in Probability and Statistics. Edited by G. del Pino and R. Rebolledo. V, 491 pages. 1986.

Vol. 1216: J. Kogan, Bifurcation of Extremals in Optimal Control. VIII, 106 pages. 1986.

Vol. 1217: Transformation Groups. Proceedings, 1985. Edited by S. Jackowski and K. Pawalowski. X, 396 pages. 1986.

Vol. 1218: Schrödinger Operators, Aarhus 1985. Seminar. Edited by E. Balslev. V, 222 pages. 1986.

Vol. 1219: R. Weissauer, Stabile Modulformen und Eisensteinreihen. III, 147 Seiten. 1986.

Vol. 1220: Séminaire d'Algèbre Paul Dubreil et Marie-Paule Malliavin. Proceedings, 1985. Edité par M.-P. Malliavin. IV, 200 pages. 1986.

Vol. 1221: Probability and Banach Spaces. Proceedings, 1985. Edited by J. Bastero and M. San Miguel. XI, 222 pages. 1986.

Vol. 1222: A. Katok, J.-M. Strelcyn, with the collaboration of F. Ledrappier and F. Przytycki, Invariant Manifolds, Entropy and Billiards; Smooth Maps with Singularities. VIII, 283 pages. 1986.

Vol. 1223: Differential Equations in Banach Spaces. Proceedings, 1985. Edited by A. Favini and E. Obrecht. VIII, 299 pages. 1986.

Vol. 1224: Nonlinear Diffusion Problems, Montecatini Terme 1985. Seminar. Edited by A. Fasano and M. Primicerio. VIII, 188 pages. 1986.

Vol. 1225: Inverse Problems, Montecatini Terme 1986. Seminar. Edited by G. Talenti. VIII, 204 pages. 1986.

Vol. 1226: A. Buium, Differential Function Fields and Moduli of Algebraic Varieties. IX, 146 pages. 1986.

Vol. 1227: H. Helson, The Spectral Theorem. VI, 104 pages. 1986.

Vol. 1228: Multigrid Methods II. Proceedings, 1985. Edited by W. Hackbusch and U. Trottenberg. VI, 336 pages. 1986.

Vol. 1229: O. Bratteli, Derivations, Dissipations and Group Actions on C^*-algebras. IV, 277 pages. 1986.

Vol. 1230: Numerical Analysis. Proceedings, 1984. Edited by J.-P. Hennart. X, 234 pages. 1986.

Vol. 1231: E.-U. Gekeler, Drinfeld Modular Curves. XIV, 107 pages. 1986.

Vol. 1232: P.C. Schuur, Asymptotic Analysis of Soliton Problems. VIII, 180 pages. 1986.

Vol. 1233: Stability Problems for Stochastic Models. Proceedings, 1985. Edited by V.V. Kalashnikov, B. Penkov and V.M. Zolotarev. VI, 223 pages. 1986.

Vol. 1234: Combinatoire énumérative. Proceedings, 1985. Edité par G. Labelle et P. Leroux. XIV, 387 pages. 1986.

Vol. 1235: Séminaire de Théorie du Potentiel, Paris, No. 8. Directeurs: M. Brelot, G. Choquet et J. Deny. Rédacteurs: F. Hirsch et G. Mokobodzki. III, 209 pages. 1987.

Vol. 1236: Stochastic Partial Differential Equations and Applications. Proceedings, 1985. Edited by G. Da Prato and L. Tubaro. V, 257 pages. 1987.

Vol. 1237: Rational Approximation and its Applications in Mathematics and Physics. Proceedings, 1985. Edited by J. Gilewicz, M. Pindor and W. Siemaszko. XII, 350 pages. 1987.

Vol. 1238: M. Holz, K.-P. Podewski and K. Steffens, Injective Choice Functions. VI, 183 pages. 1987.

Vol. 1239: P. Vojta, Diophantine Approximations and Value Distribution Theory. X, 132 pages. 1987.

Vol. 1240: Number Theory, New York 1984–85. Seminar. Edited by D.V. Chudnovsky, G.V. Chudnovsky, H. Cohn and M.B. Nathanson. V, 324 pages. 1987.

Vol. 1241: L. Gårding, Singularities in Linear Wave Propagation. III, 125 pages. 1987.

Vol. 1242: Functional Analysis II, with Contributions by J. Hoffmann-Jørgensen et al. Edited by S. Kurepa, H. Kraljević and D. Butković. VII, 432 pages. 1987.

Vol. 1243: Non Commutative Harmonic Analysis and Lie Groups. Proceedings, 1985. Edited by J. Carmona, P. Delorme and M. Vergne. V, 309 pages. 1987.

Vol. 1244: W. Müller, Manifolds with Cusps of Rank One. XI, 158 pages. 1987.

Vol. 1245: S. Rallis, L-Functions and the Oscillator Representation. XVI, 239 pages. 1987.

Vol. 1246: Hodge Theory. Proceedings, 1985. Edited by E. Cattani, F. Guillén, A. Kaplan and F. Puerta. VII, 175 pages. 1987.

Vol. 1247: Séminaire de Probabilités XXI. Proceedings. Edité par J. Azéma, P.A. Meyer et M. Yor. IV, 579 pages. 1987.

Vol. 1248: Nonlinear Semigroups, Partial Differential Equations and Attractors. Proceedings, 1985. Edited by T.L. Gill and W.W. Zachary. IX, 185 pages. 1987.

Vol. 1249: I. van den Berg, Nonstandard Asymptotic Analysis. IX, 187 pages. 1987.

Vol. 1250: Stochastic Processes – Mathematics and Physics II. Proceedings 1985. Edited by S. Albeverio, Ph. Blanchard and L. Streit. VI, 359 pages. 1987.

Vol. 1251: Differential Geometric Methods in Mathematical Physics. Proceedings, 1985. Edited by P.L. García and A. Pérez-Rendón. VII, 300 pages. 1987.

Vol. 1252: T. Kaise, Représentations de Weil et GL_2 Algèbres de division et GL_n. VII, 203 pages. 1987.

Vol. 1253: J. Fischer, An Approach to the Selberg Trace Formula via the Selberg Zeta-Function. III, 184 pages. 1987.

Vol. 1254: S. Gelbart, I. Piatetski-Shapiro, S. Rallis. Explicit Constructions of Automorphic L-Functions. VI, 152 pages. 1987.

Vol. 1255: Differential Geometry and Differential Equations. Proceedings, 1985. Edited by C. Gu, M. Berger and R.L. Bryant. XII, 243 pages. 1987.

Vol. 1256: Pseudo-Differential Operators. Proceedings, 1986. Edited by H.O. Cordes, B. Gramsch and H. Widom. X, 479 pages. 1987.

Vol. 1257: X. Wang, On the C^*-Algebras of Foliations in the Plane. V, 165 pages. 1987.

Vol. 1258: J. Weidmann, Spectral Theory of Ordinary Differential Operators. VI, 303 pages. 1987.

continued on page 205

PREFACE

The Twelfth Summer Symposium in Real Analysis, August 9-12 1988, was held at the Coleraine campus of the University of Ulster. The theme of the conference was new integrals and related subjects, with particular reference to the generalised Riemann integral of R. Henstock and J. Kurzweil. This volume contains a brief introduction to the theme by R. Henstock, and a survey article by S. Foglio.

In holding the conference, the support of the following is acknowledged: The Royal Society, London Mathematical Society, Royal Irish Academy, Institute of Mathematics and its Applications, Department of Education (UK), Editorial Board of Real Analysis Exchange, University of Ulster, Ulster Bank, the Bushmill's Distillery and the Northern Ireland Tourist Board.

The editors express their gratitude to Mr. Gerry Shannon, Mathematics Department, University of Ulster, for his contribution to the production of these proceedings.

In addition, thanks are due to Frank Cassidy, Mike Davison, Joan Shannon, Máire McKeever, Leanne Woods and Chris Butler for their help.

P. M.

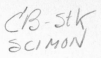

CONTENTS

About Ralph Henstock

Pat Muldowney

The mining village of Newstead, near Nottingham, is in the parish of St. Mary the Virgin in Sherwood Forest, one of three parishes said to have been created by King Henry II in penance for the murder of Thomas à Becket; the other two being York Minster and Westminster Abbey. The Priory founded there was, in a later era, the ancestral seat of the poet Byron.

Ralph Henstock was born in Newstead on 2nd June 1923, to William Henstock, a mineworker and former miner, and Mary Ellen Henstock (née Bancroft). His father was determined that Ralph would never go down the mine, and, as a gifted young scholar, he grew up immersed in the popular science of the time.

It was assumed that he would attend Nottingham University, where both his father and his uncle had acquired technical training, but Henstock wished to go to the university of Isaac Newton. For some time he concentrated on chemistry, but switched to mathematics when he discovered, to his surprise, that this could be taken as a degree subject.

In 1941 he took the London Higher Schools Examination (equivalent to the baccalaureate or A-Levels) in double Mathematics, including a scholarship component, plus Physics. He got a "Good" in Physics and a distinction in every paper of Mathematics and was awarded a State Scholarship. He also gained a Nottingham County and Kesteven (district of Lincolnshire) Senior Scholarship and the Henry Mellish Scholarship (given to the top candidate in Nottingham City and County in the same examinations).

With these three scholarships, and with a required pass in Latin (obtained by intensive cramming over eight weeks!) in the Cambridge Previous Examination, he entered St. John's College, Cambridge, in October 1941.

Because of war-time conditions his class was allowed only two years of undergraduate study, so he was placed in the second year of the Mathematics Tripos, omitting the one-year Part I. Very little of the Cambridge frivolities remained, apart from rowing, which he avoided, and the Archimedeans (Cambridge Mathematical Society). Among the people he came into contact with were such contrary spirits as J.D.Bernal, advocate of mathematics for use, and G.H.Hardy, who believed mathematics contains its own justification.

In 1942-43 he attended a set of lectures by J.C. Burkill on Lebesgue integration, as an extra subject for the Cambridge Part II Tripos. He was classified

Wrangler in Part II and so was made scholar of St. John's in 1943. In 1942-43 two courses of lectures in statistics were added to the curriculum, and in November 1943 Henstock was sent to London to the Ministry of Supply Advisory Service on Statistical Method and Quality Control.

He soon tired of statistics and joined Birkbeck, apparently the only college left in London, where he took various courses on summability and integration. He did a great deal of research in London's Patents Library, the only one he could find, at that time, containing mathematical publications.

C.A. Rogers continues the narrative :

"Once-upon-a-time, a young mathematician called on Professor Paul Dienes at Birkbeck College in London. He asked to be accepted as a postgraduate student and mentioned that he would like to try to modify a result on the theory of divergent series to obtain a result on divergent integrals. Dienes replied - No! you shall work on the theory of integration. Mr Henstock, as he then was, agreed to this suggestion – it was really a suggestion rather than a command. Although Dienes could not prevent Henstock from working from time to time on divergent series we must be thankful to him for ensuring that Henstock has in the main concentrated on the theory of integration.

"At that time, Dienes ran a weekly seminar at Birkbeck College, while Davenport ran one at University College. These were the foci of mathematical activity in London. Dienes was assisted by the boundless enthusiasm of R.G. Cooke. I first met Henstock at this seminar. It was frequented by J.L.B. Cooper; Abraham Robinson frequently attended, and John Todd, always accompanied by Olgar Tausky, occasionally turned up. It was a very stimulating affair".

Such was the milieu in which Henstock began his mathematical work.

In 1944 he was awarded the Cambridge B.A. and then began research for the Ph. D. in London, which he gained in December 1948. His thesis dealt with summability theory, integration theory and Walsh functions. With reference to the latter, he invented independently [6] a table of signs that defined the functions which had been discovered earlier by J.L.Walsh. The examiners were J.C. Burkill and H. Kestelman.

Having acquired a lifelong aversion to the Civil Service, he was released from the Ministry of Supply in January 1947 and went back to Cambridge for the next two terms to take Part III of the Mathematics Tripos. Returning to London as an Assistant at Bedford College for Women, Regent's Park 1947-48, he then became an Assistant Lecturer at Birkbeck, 1948-51. His subsequent academic career is, in outline:

Lecturer in Pure Mathematics at Queen's University, Belfast 1951-56,
Lecturer at Bristol University 1956-60,
Senior Lecturer, Queen's, Belfast 1960-62,
Reader, Queen's, 1962-64,

Reader, Lancaster University 1964-70,

Chair, New University of Ulster, University of Ulster 1970-88,

Emeritus Professor (Leverhulme Fellowship), Ulster, 1988-.

Broadly speaking Henstock's mathematical work is in the areas of summability, linear analysis, and integration theory, and he has made substantial contributions in all these fields. From time to time he has also worked in statistics, looking for a better theory of significance tests.

The contribution with which Henstock is particularly associated is the generalised Riemann or Riemann complete integral which features in his work from the early fifties. See, for instance, his use of gauge functions in [9] pp. 277-278, published in 1955. The first comprehensive formulation of a theory of integration based on gauge functions [20] appears in 1963.

Since 1962 he has had a particular interest in the problems of integration in function spaces, and this has apparently motivated a general conception of integration, of which the gauge integral is an aspect, and which he continues to develop extensively.

"Many mathematicians obtain striking results or beautiful results and some even obtain strikingly beautiful results. But very few obtain impossible results that actually turn out to be possible. Let me explain by giving two examples. It was in 1892 that E.G. Peano constructed his space-filling curve. This was completely counter to all previous intuitive ideas. Mathematicians had always thought that a curve was a curve and was quite different from a surface. But here was a curve that actually was a surface. This single discovery led inevitably to a new mathematical subject called Dimension Theory.

"My second example is the discovery independently by Kurzweil and Henstock of a new method of integration. The first major method of integration, after Newton's work, was discovered by Riemann. The second method due to Lebesgue was in many ways a striking improvement leading to a most beautiful theory. Such were the advantages of Lebesgue's theory that mathematicians firmly believed that no modification of Riemann's method could possibly give such powerful results.

"In their first papers Kurzweil and Henstock place no emphasis on the fact that they had introduced a Riemann-type integral that was as powerful as Lebesgue's. The distinguished reviewers of these papers in Mathematical Reviews did not pick up this point. It was only when Henstock published his book and Hildebrant published a generous and scholarly review in Mathematical Reviews that mathematicians woke up to the fact that the impossible had happened and that here was a Riemann-type integral that was as good as, and indeed, in many ways far superior to Lebesgue's. A great rejuvenation and flowering of integration theory resulted." [1]

In 1949 Henstock married Marjorie Jardine in Wesley's Chapel, London. Their son John was born in 1952 in Belfast. A devoted family man, he takes an active interest in community affairs in Northern Ireland where he has spent most of his working life. Often accompanied by his dog Becky, Henstock is a familar figure on the Coleraine campus of the University of Ulster. He combines in his personality great individuality, integrity and conviction with equally great human sympathy, humour and consideration. He has set very high standards for his students, but is unstinting in his support and assistance to enable them to meet these standards. His main interests outside of mathematics are in Methodism and New Testament Christianity, and also in poetry.

The Twelfth Summer Symposium in Real Analysis was held at Henstock's home university of Coleraine in recognition of his outstanding contribution to this subject.

List of publications of R. Henstock.

C = Canadian Journal of Mathematics
I = Proceedings Royal Irish Academy, series A
J = Journal London Mathematical Society
MR = Mathematical Reviews
P = Proceedings London Mathematical Society

1. *On interval functions and their integrals*, J 21 (1946) 204-09, MR 8, 572.

2. *The efficiency of matrices for Taylor series*, J 22 (1947) 104-07, MR 9, 278.

3. *On interval functions and their integrals (II)*, J 23 (1948) 118-28, MR 10, 239.

4. *The efficiency of matrices for bounded sequences*, J 25 (1950) 27-33, MR 11, 429.

5. *Sets of uniqueness for trigonometric series and integrals*, Proceedings Cambridge Philosophical Society, 46 (1950) 538-48, MR 12, 496.

6. *Density integration*, P (2) 53 (1951) 192-211, MR 13, 20.

7. (with A. M. Macbeath) *On the measure of sum-sets (I) The theorems of Brunn, Minkowski, and Lusternik*, P (3) 3 (1953) 182-94, MR 15, 109.

8. *Linear functions with domain a real countably infinite dimensional space*, P (3) 5 (1955) 238-56, MR 17, 176.

9. *The efficiency of convergence factors for functions of a continuous real variable*, J 30 (1955) 273-86, MR 17, 359.

10. *Linear and bilinear functions with domain contained in a real countably infinite dimensional space*, P (3) 6 (1956) 481-500, MR 18, 584.

11. *On Ward's Perron-Stieltjes integral*, C 9 (1957) 96-109, MR 18, 880.

12. *The summation by convergence factors of Laplace-Stieltjes integrals outside their half-plane of convergence*, Mathematische Zeitschrift 67 (1957) 10-31, MR 18, 880.

13. *A new descriptive definition of the Ward integral*, J 35 (1960) 43-48, MR 22 #1648.

14. *The use of convergence factors in Ward integration*, P (3) 10 (1960) 107-21, MR 22 #12197.

15. *The equivalence of generalized forms of the Ward, variational, Denjoy-Stieltjes, and Perron-Stieltjes integrals*, P (3) 10 (1960) 281-303, MR 22 #12198.

16. *N-variation and N-variational integrals of set functions*, P (3) 11 (1961) 109-33, MR 23 #A995.

17. *Definitions of Riemann type of the variational integrals*, P (3) 11 (1961) 402-18, MR 24 #A1994.

18. *Difference-sets and the Banach-Steinhaus theorem*, P (3) 13 (1963) 305-21, MR 26 #6776.

19. *Tauberian theorems for integrals*, C 15 (1963) 433-39, MR 28 #2187.

20. Theory of integration, (Butterworths, London, 1963), MR 28 #1274.

21. *The integrability of functions of interval functions*, J 39 (1964) 589-97 MR 29 #5975.

22. *Majorants in variational integration*, C 18 (1966) 49-74, MR 32 #2545.

23. *A Riemann-type integral of Lebesgue power*, C 20 (1968) 79-87, MR 36 #2754.

24. Linear Analysis, (Butterworths, London, 1968), MR 54 #7725.

25. *Generalized integrals of vector-valued functions*, P (3) 19 (1969) 509-36, MR 40 #4420.

26. *Integration by parts*, Aequationes Math. 9 (1973) 1-18, MR 47 #3608.

27. (with S. Foglio) *The N-variational integral and the Schwartz distributions III*, J (2) 6 (1973) 693-700, MR 48 # 2755.

28. *Integration in product spaces, including Wiener and Feynman integration*, P (3) 27 (1973) 317-44, MR 49 #9145.

29. *Additivity and the Lebesgue limit theorems*, The Greek Mathematical Society C. Caratheodory Symposium Sept. 3-7, 1973, pp. 223-41 (published 1974), MR 57 #6355.

30. *Integration, variation and differentiation in division spaces*, I 78 (1978) No. 10, 69-85, MR 80d: 26011.

31. *The variation on the real line*, I 79 (1979) No. 1, 1-10, MR 81d: 26005.

32. *Generalized Riemann integration and an intrinsic topology*, C 32 (1980), 395-413, MR 82b: 26010.

33. *Division spaces, vector-valued functions and backwards martingales*, I 80 (1980), No. 2, 217-232, MR 82i: 60091.

34. *Density integration and Walsh functions*, Bulletin Malaysian Math. Soc. (2) 5 (1982) 1-19.

35. *A problem in two-dimensional integration*, Journal Australian Math. Soc. (series A) 35 (1983) 386-404, MR84k: 26010.

36. *The Lebesgue syndrome*, Real Analysis Exchange 9 No. 1 (1983-84) 96-110.

37. *The reversal of power and integration*, Bulletin Institute of Math. and App. 22 (3-4) (1986) 60-61.

38. Lectures on the theory of integration, *World Scientific*, Singapore, 1988.

39. *A short history of integration theory*, Southeast Asian Bulletin of Math. 12, No.2 (1988).

40. General theory of integration, (Oxford University Press - in preparation).

Introduction to the new integrals

Ralph Henstock

[1] The constructed definite integral of most calculus books, originally developed by Leibnitz, is now that of Riemann [18] and Darboux [6]. In 1902 its limitations were highlighted by Lebesgue's [13] publication of his own integral. In particular, the taking of limits under the integral sign was rendered much more straightforward by Lebesgue's integral. The 87 years since that famous paper have seen some very simple constructions of the Lebesgue integral but none to approach the simplicity of the Riemann integral until the Riemann-complete (or generalized Riemann or gauge) integral was developed in the late 1950's.

Paradoxically, the gauge integral has a simpler definition than the Lebesgue integral and yet covers a wider field, being equivalent to Denjoy's [7] totalisation process of 1912, also called the special Denjoy (or Denjoy-Perron) integral. This was an extension of Lebesgue's integral using two processes repeatedly in a transfinite induction, and in its original form was a rather complicated construction to obtain the primitive (or integrated function) from a given ordinary derivative, however complicated it is.

Again, the intervening 77 years have produced many simplifications, notably by Lusin [14] and Perron [17], but none as simple, practical and direct as the gauge integral of J.Kurzweil [12] and R.Henstock [8, 11]. The fixed norm or mesh of Riemann integration (in one dimension, the maximum length of intervals of a division) becomes a function called a gauge in gauge integration, and it is as if the Lebesgue decades from 1902 to 1955 have all been swept away, and we are just developing the calculus constructed integral still further.

To obtain the Lebesgue integral of f we need only use the modification of McShane [16], or say that f and $|f|$ are gauge integrable. The gauge integral takes the Stieltjes [19] integral, $\int f dg$, and Burkill's [1] integral $\int h$ of interval and rectangle functions in its stride.

The gauge integral is just one of a variety of integrals constructed in a similar way using division spaces, most of which have the Arzelà-Lebesgue limit properties and Lebesgue differentiation properties. In many cases where a function oscillates unboundedly and too badly to be gauge integrable, it is sometimes possible to smooth the oscillations and so have Burkill's Cesàro-Perron integrals [2,3,4,5], the Marcinkiewicz-Zygmund integral [15], and the

[1] **This introduction was written during the term of a Leverhulme Research Fellowship.**

N-variational integrals [9,10].

This is the general field of integration theory in which the papers of this Symposium can be embedded.

References

1. J.C.Burkill, *Functions of intervals*, Proc. London Math Soc. (2) 22 (1924) 275-310.
2. J.C.Burkill, *The Cesàro-Perron integral*, Proc. London Math. Soc. (2) 34 (1932) 314-322, Zbt.5.392.
3. J.C.Burkill, *The Cesàro-Perron scale of integration*, Proc. London Math. Soc. (2) 39 (1935) 541-552, Zbt.12.204.
4. J.C.Burkill, *Fractional orders of integrability*, Journal London Math. Soc. 11 (1936) 220-226, Zbt.14.258.
5. J.C.Burkill, *Integrals and trigonometric series*, Proc. London Math. Soc. (3) 1 (1951) 46-57, MR13, 126.
6. J.G.Darboux, *Mémoire sur les fonctions discontinues*,Ann. Sci. Ec. Norm. Sup. (2) 4 (1875) 57-112.
7. A.Denjoy, *Une extension de l'intégrale de M.Lebesgue*, C.R.Acad. Sci. Paris 154 (1912) 859-862.
8. R.Henstock, *The efficiency of convergence factors for functions of a continuous real variable*, Journal London Math. Soc. 30 (1955) 273-286, MR17, 359.
9. R.Henstock, *The use of convergence factors in Ward integration*, Proc. London Math. Soc. (3) 10 (1960) 107-121, MR22#12197.
10. R.Henstock, *The equivalence of generalized forms of the Ward, variational, Denjoy-Stieltjes and Perron-Stieltjes integrals*, Proc. London Math. Soc. (3) 10 (1960) 281-303, MR22#12198.
11. R.Henstock, *Definitions of Riemann type of the variational integrals*, Proc. London Math. Soc. 11 (1961) 402-418, MR24#A1994.
12. J.Kurzweil, *Generalized ordinary differential equations and continuous dependence on a parameter*, Czech. Math. Journal 7 (82) (1957) 418-446 (especially 422-428), MR22#2735.
13. H.Lebesgue, *Intégrale, languer, aire*, Annali Mat. Pura Appl. (3) 7 (1902) 231-359, J.buch 33, 307.

14. N.Lusin, ***Sur les propriétés de l'intégrale de M.Denjoy***, C.R.Acad. Sci. Paris 155 (1912) 1475-1478.

15. J.Marcinkiewicz and A.Zygmund, ***On the differentiability of functions and summability of trigonometrical series***, Fundamenta Math. 26 (1936) 1-43, Zbt.14.111.

16. E.J.McShane, ***A unified theory of integration***, American Math. Monthly 80 (1973) 349-359, MR47#6981.

17. O.Perron, ***Ueber den Integralbegriff***, S.-B. Heidelberg Akad. Wiss., Abt.A16 (1914) 1-16.

18. G.F.B.Riemann, ***Über die Darstellbarkeit einer Funktion durch eine trigonometrische Reihe***, Abh. Gesell. Wiss. Göttingen 13 (1868) Math. Kl. 87-132, Oeuvres mathématiques de Riemann, 1898, Paris, reprinted 1968, Paris and Cleveland.

19. T.J.Stieltjes, ***Recherches sur les fractions continues***, Annales de la Faculté des Sci. de Toulouse 8 (1894).

Department of Mathematics
University of Ulster
Coleraine
Co. Londonderry
Northern Ireland

Some applications of a theorem of Marcinkiewicz

P.S. Bullen

Contents:

1. THE THEOREM

1.1 Let $f : [a, b] \to \mathbb{R}$. Then to say that M is a *major function* of f, $M \in \overline{M}(f)$, means that $(\overline{1})$:

$$M : [a, b] \to \mathbb{R};$$

$$M(a) = 0;$$

$$\underline{D}M \geq f.$$

If $-m$ is a major function of $-f$ then we say that m is a *minor function* of f, $m \in \underline{M}(f)$; equivalently, $(\underline{1})$:

$$m : [a, b] \to \mathbb{R};$$

$$m(a) = 0;$$

$$\overline{D}M \leq f.$$

If $M \in \overline{M}(f)$ and $m \in \underline{M}(f)$ we will call (M, m) a *bounding pair (of bounding functions)* for f. If such a pair exists and if

$$\inf\{M(b);\ M \in \overline{M}(f)\} = \sup\{m(b);\ m \in \underline{M}(f)\} = I,$$

then f is said to be ***Perron integrable*** on $[a, b]$, $f \in P^*(a, b)$, (or just $f \in P^*$ if $[a, b]$ is understood); I is called the ***P^*-integral*** of f on $[a, b]$, written $P^* \int_a^b f$, (or just $\int_a^b f$ if there is no ambiguity).

The standard elementary theory of an integral, for the P^*-integral, is, with a few exceptions, easily deduced, (see Saks [1, p. 190]). An essential tool is the fact that if (M, m) is any bounding pair then $M - m$ is increasing. In particular then the P^*-*primitive* of an $f \in P^*$ is defined:

$$F(x) = \int_a^x f, \quad a \le x \le b;$$

F is continuous and $F' = f$ a.e. (This of course implies that if $f \in P^*$ then f is measurable).

(It is convenient to note that if M is a super-additive interval function defined on the closed sub-intervals of $[a, b]$ satisfying

$$M(I) \ge f(x)|I|,$$

($|I|$=length of I), for all $x \in [a, b]$ and all closed intervals I, $I =]x - \delta, x + \delta[$, $\delta = \delta(x) > 0$, then $M(x) = M([a, x])$, $a \le x \le b$, is a major function. M will denote both the point and interval functions related in this way).

1.2. We can now state the theorem that is the subject of this talk.

Theorem 1. If $f : [a, b] \to \mathbb{R}$ is measurable and if f has at least one continuous bounding pair then $f \in P^*(a, b)$.

This is a remarkable theorem with surprisingly few hypotheses. The first condition of measurability is not implied by the others and is needed for the conclusion, as we saw above; it will not be discussed any further.

The continuity conditions seems on the other hand a strong requirement. However the original P^*-integral theory was developed using continuous bounding functions; see Perron [1], Bauer [1]. It was only Saks in 1937 who proved that the same theory resulted without this restriction. So classically continuity was included in the concept of bounding pair; and it is the definition of bounding functions that is the implicit third hypothesis in Theorem 1.

An advantage of the classical approach is that we can generalise the concept of bounding functions. Thus $(\bar{1})$ can be replaced by $(\bar{1})_g$

$$M : [a, b] \to \mathbb{R};$$

$$M(a) = 0;$$

$$\underline{D}M \ge f, \text{ a.e.};$$

$$\underline{D}M > -\infty, \text{ n.e.};$$

(n.e.= nearly everywhere = except on a countable set.) If then we make a similar change in $(\underline{1})$, $(\underline{1}_g)$ say, and if (M, m) is a generalised bounding pair, then $M - m$ is still increasing if M and m are taken to be continuous; see Bruckner [1, p. 189]. That continuous generalized bounding functions give the same theory as continuous bounding functions is a classical result; see Bauer [1].

1.3. The last remark, while not trivial is not very difficult to prove. On the other hand the result of Saks is of completely different type and depends on one of the deep theorems of the whole theory: $f \in P^*$ ***iff there exists a continuous ACG^* function F such that $F' = f$ a.e.***

In fact the Marcinkiewicz theorem is not really a result in the theory of the Perron integral. Rather it is a property of the D^*-integral of Denjoy that is defined by the above theorem, often called the Saks-Hake-Looman-Aleksandrov theorem, Saks [1; pp. 247-252]. Then the above theorem is just $f \in P^*$ iff $f \in D^*$ and Theorem 1 can be given a stronger form:

Theorem 2. If $f : [a, b] \to I\!R$ is measurable then $f \in D^*(a, b)$ iff f has at least one continuous bounding pair.

1.4. Theorem 1, attributed to J. Marcinkiewicz, appears in the second edition of Saks' book, Saks [1, p. 253], published late in 1937. As far as I know Marcinkiewicz did not publish this result elsewhere.

In 1939 G. Tolstov [1] proved the same result. Presumably Tolstov did not know of the Saks' book as it was not published in Moscow until 1949. His proof is the same as that of Marcinkiewicz; it takes up more space because several useful lemmas, already available in Saks' book, have to be developed.

In 1949 A. Denjoy published a deep and exhaustive discussion of the Perron approach to his integral, Denjoy [1, pp. 616-683]. In the course of this he also proves the Marcinkiewicz theorem, (in a slightly more general form that will be mentioned later); the proof only appears to be different because it is given within the Denjoy theory of variation and totalization.

1.5. We now consider generalizing the conditions of Theorem 1. In his book, Saks [1, p.253], Saks suggests that the continuity condition is essential, giving the following example.

Example 1.

$$\begin{aligned} f(x) &= -x^{-2}, && 0 < x \le 1, \\ &= 0, && x = 0; \\ M(x) &= 0, && 0 \le x \le 1; \\ m(x) &= x^{-1}, && 0 < x \le 1; \\ &= 0, && x = 0. \end{aligned}$$

Then $f \notin P^*(0, 1)$; $M \in C(0,1)$, but $m \notin C(0,1)$.

In 1978 Sarkhel [1] pointed out that Example 1 is not convincing, in that m is not a minor function in the sense of (1), since there is a point at which $\bar{D}m(x) \le f(x)$ fails to hold. Example 1 does indicate that Theorem 3 may fail if the continuity condition is dropped, although even here we might remark that $M - m$ is not increasing so in fact we should not be looking on (M, m) as a bounding pair for f. A more convincing example is the following:

Example 2.

$$\begin{aligned}
\phi(x) &= x^2 \sin(1/x), && 0 < x \le 1;\\
&= 0, && x = 0;\\
f(x) &= \phi''(x), && 0 < x \le 1;\\
&= 0, && x = 0;\\
M(0) &= m(0) = 0; &&\\
M(x) &= 1 + \phi'(x), \quad m(x) = -1 + \phi'(x), && 0 < x \le 1.
\end{aligned}$$

Then $f \notin P^*(0,\ 1)$; M and m are discontinuous at the origin, although M is lower-, and m is upper-, semicontinuous.

Sarkhel gave a generalization of Theorem 3; in its simplest form it gives the following generalization of Theorem 1.

Theorem 3. If $f :\ [a,\ b] \to I\!R$ is measurable and if f has at least one regulated bounding pair then $f \in P^*(a,\ b)$.

An inspection of the proof of Theorem 1 shows that, given a measurable $f :\ [a,\ b] \to I\!R$ then $f \in P^*(a,\ b)$ provided we can find a pair of functions (U, L) defined on $[a, b]$, (and with $U(a) = L(a) = 0$), that satisfies the following conditions. We will only give the conditions needed for the upper function U, those for the lower function L are analogous.

(1) U is BVG*; (this holds in particular if $\underline{D}U > -\infty$ n.e.);

(2) on the set where U' is defined, (and so, by (1), a.e.), $U' \ge f$;

(3) for all $\varepsilon > 0$, x, $a \le x \le b$, there is a $\delta = \delta(x,\ \varepsilon) > 0$ such that if I is a closed interval and $I = [x - \delta,\ x[$, or $I =]x,\ x + \delta[$ then $U(I) \le \varepsilon$;

(4) on any interval I on which f is P^*-integrable $U(I) \ge F(I)$, where F is a P^*-primitive of f.

This last condition is equivalent to the following; let $\underline{M}^*(f)$ be any class of minor functions that can be used to define the P^*-integral;

(4)* for all $m \in \underline{M}^*(f)$, $U - m$ is increasing.

Briefly: conditions (1) and (2) are used to obtain the Harnack property of the P^*-integral; conditions (3) and (4) are used to prove the Cauchy property of the P^*-integral.

A consideration of these conditions, and the way they are used, enabled Bullen and Vyborny [1] to prove the following generalization of Theorem 1; the continuity condition is dropped but more thsn one bounding pair is allowed.

Theorem 4. Hypotheses: $f :\ [a,\ b] \to I\!R$ is measurable; $\bar{A} \subset \overline{M}(f)$, $\bar{A} \ne \emptyset$, $\underline{A} \subset \underline{M}(f)$, $\underline{A} \ne \emptyset$, for all $\varepsilon > 0$, x, $a \le x \le b$, there is a $\delta = \delta(x,\ \varepsilon) > 0$ and an $M \in \bar{A}$, $m \in \underline{A}$, such that if I is a closed interval, $I \subset]x - \delta,\ x[$, or $I \subset]x,\ x + \delta[$, then $-\varepsilon \le m(I) \le M(I) \le \varepsilon$. Conclusion: $f \in P^*(a,\ b)$.

This generalizes Theorem 1 for if (M, m) is a regulated bounding pair we can take $\bar{A} = \{M\}$, $\underline{A} = \{m\}$; in general of course the pair (M, m) depends on x and ε.

2. APPLICATIONS

2.1. Equivalence of Certain Integrals. Suppose that in $(\bar{1})$ we replace $\underline{D}M \geq f$ by $\underline{D}_+M \geq f$, and make a similar change in $(\underline{1})$ then a Perron integral can be defined; see Ridder [2]. Let us call such a pair (M, m) a ***right bounding pair*** and the integral the P_+^*-integral. Clearly if $f \in P^*(a, b)$ then $f \in P_+^*(a, b)$: in fact the integrals are equivalent as is seen from the following generalization of Theorem 1 due to McShane [1, Lemma 5.2].

Theorem 5. If $f : [a, b] \to \mathbb{R}$ is measurable and if f has at least one continuous right bounding pair then $f \in P^*(a, b)$.

Of the conditions in 1.5 only (1) needs any consideration since $\underline{D}_+M > -\infty$ only implies M is BVG; M is not necessarily BVG*; see Saks [1. p. 237]; (in particular Lemma 5.1 of McShane [1] is incorrect). However given a right bounding pair (M, m) the following inequalities hold:

$$\underline{D}_+M \geq \overline{D}_+m,\ \underline{D}_+M > -\infty,\ \overline{D}_+m < \infty;$$

and they do imply that both M and m are BVG* see Ridder [2], or Sarkhel [1]. It might be noted that this theorem can be generalized further by requiring only that

$$\overline{D}_+M \geq f \geq \underline{D}_+m,\ \underline{D}_+M > -\infty,\ \overline{D}_+m < \infty;$$

see Denjoy [1].

2.2. Convergence Theorems. A major difficulty in applying the P^*-integral to analysis has been the lack of a good convergence theorem; one that is not really a Lebesgue theory result. Thus if $f \leq g_n \leq h$, f, g_n, h all P^*-integrable, $n \in \mathbb{N}$; then $0 \leq g_n - f \leq h - f$, $n \in \mathbb{N}$, and these differences, being non-negative, are L-integrable and so the standard theory of dominated convergence can be developed.

The basic convergence theorem for P^*-integrals is due to Džvaršeĭšvili see Čelidze and Džvaršeĭšvili [1, pp.47-54]. It starts from the following result of Vitali; see Natanson [1, p.152].

Theorem 6. Hypotheses: $f_n \in L(a, b)$, $F_n(x) = L\int_a^x f_n$, $a \leq x \leq b$, $n \in \mathbb{N}$, $\lim_{n\to\infty} f_n = f$ a.e.; $\{F_n,\ n \in \mathbb{N}\}$ is uniformly absolutely continuous, (UAC).

Conclusions: $f \in L(a, b)$ and $L\int_a^b f = \lim_{n\to\infty} L\int_a^b f_n$.

The idea of Džvaršeĭšvili was to replace L by P^* and UAC by UACG*. The same idea was thought of independently, and in a different setting by Lee and Chew [1]. Since the topic of convergence theorem is to be covered in another presentation, Lee [1], we mention only one of the many results of Lee and Chew; one that uses the ideas being discussed here.

Theorem 7. Hypotheses: $f_n \in P^*(a, b)$, $F_n(x) = \int_a^x f$, $a \leq x \leq b$, $n \in \mathbb{N}$; $\lim_{n\to\infty} f_n = f$; $\lim_{n\to\infty} F_n = F$, uniformly; there is at least one common bounding pair for all f_n, $n \in \mathbb{N}$.

Conclusions: $f \in P^*(a, b)$ and $\int_a^b f = F(b)$.

Here it is important that the common pair not be continuous, (or even regulated). For if this were so then by Theorem 1, (or Theorem 3), the functions $\inf f_n$, $\sup f_n$ would be P^*-integrable; then Theorem 7 would just be the dominated convergence theorem, as was remarked above.

The common bounding pair satisfies, relative to f, all of the conditions of 1.5 except (3), the continuity condition. The uniform convergence of the primitives, F_n, $n \in \mathbb{N}$, enables us to avoid using this condition; see Lee and Chew [1]; in fact we can replace uniform convergence by pointwise convergence to a continuous function, Chew [1].

2.3. Locally Small Riemann Sums. Recently Schurle [1] gave a very interesting necessary and sufficient condition for P^*-integrability. A simple proof of this result can be given, using Theorem 5; see Bullen and Vyborny [1] and Lee and Lu [1].

The Schurle condition needs the Riemann theory of the Denjoy integral, a theory due to Henstock, and, independently, to Kurzweil. Let $\delta : [a, b] \to \mathbb{R}$, $\delta > 0$, then $\pi = \{a_0, \ldots, a_n;\ y_0, \ldots, y_n\}$ is called a ***δ-fine partition*** of $[a, b]$, $\pi \in \Pi(\delta; a, b)$ if (i) $a = a_0 < \ldots, < a_n = b$, (ii) $a_{i-1} \le y_i \le a_i$, $1 \le i \le n$, (iii) $a_i - a_{i-1} < \delta(y_i)$, $1 \le i \le n$. If then $f : [a, b] \to \mathbb{R}$, and $\pi \in \Pi(\delta; a, b)$ the associated Riemann sum is $\sum_\pi f = \sum_{i=1}^n f(y_i)(a_i - a_{i-1})$. We then say f is ***R^*-integrable*** on $[a, b]$, $f \in R^*(a, b)$, if there is an I such that for all $\varepsilon > 0$ there is a $\delta > 0$ such that for all $\pi \in \Pi(\delta; a, b)$, $|\sum_\pi f - I| < \varepsilon$; then $R^* \int_a^b f = I$. It can be proved that: $f \in R^*$ iff $f \in P^*$; see Henstock [1, p. 123], Kurzweil [1, p. 37]. The result of Schurle is

Theorem 8. If $f : [a, b] \to \mathbb{R}$ is measurable then $f \in P^*(a, b)$ iff for all $\varepsilon > 0$ there is a $\delta = \delta(x, \varepsilon) > 0$ such that if $a \le c \le x \le d \le b$, $0 < d - c < \delta(x)$ then for all $\pi \in \Pi(\delta; c, d)$, $|\sum_\pi f| < \varepsilon$.

The condition in this theorem was called by Schurle locally small Riemann sums, LSRS. The existence of LSRS allows the construction of superadditive interval functions, that in turn allow us to define families $\bar{A}$, $\underline{A}$ of Theorem 4 and so prove Theorem 8.

2.4. Differential Equations. A final application is to differential equations and generalizes a classical result; see Coddington [1, p. 43]. We use the following notations:

$$I = [\tau - a,\ \tau + a];\ J = [\xi - b,\ \xi + b],$$

and if $f : I \times J \to \mathbb{R}$, and $g : I \to J$ we write $f_g : I \to \mathbb{R}$ for the function defined by

$$f_g(t) = f(t,\ g(t)),\ t \in I.$$

Theorem 9. Hypotheses: $f : I \times J \to \mathbb{R}$ is such that

(i) $f(t, 0)$ is continuous on J for almost all $t \in I$;

(ii) for all continuous ACG* $g : I \to J$, f_g is measurable;

(iii) there exist continuous M, m on I, $M(\tau) = m(\tau) = 0$, such that for all continuous and ACG* g, $g : I \to J$, $\underline{D}M \ge f_g \ge \overline{D}m$.

Conclusions. There is a continuous ACG* ϕ, $\phi : I \to I\!R$, with $\phi(\tau) = \xi$, which is a solution of the integral equation

$$x(t) = \xi + \int_\tau^t f(s, x(s))ds, \ t \in I.$$

The proof uses Theorem 1 to develop the usual approximations, but now using P^*-integration, rather than L-integration; that the approximations converge then follows by an application of Theorem 7; see Bullen and Vyborny [1].

3. GENERALIZATIONS

Before discussing whether other general integrals have a Marcinkiewicz theorem it should be noted that such a theorem exists for the Lebesgue integral. In fact Theorem 1 is a very natural generalization of the following result: *if* $f : [a, b] \to I\!R$ *is measurable and if* f *has at least one BV bounding pair then* $f \in L(a, b)$.

The most obvious question to ask is whether the general Denjoy integral, the D-integral, has a Marcinkiewicz theorem. Since for this integral the correct derivative to use is the approximate derivative we would use ***approximate bounding pairs***, obtained by replacing $\underline{D}M \geq f$ in $(\bar{1})$ by $\underline{D}_{\text{ap}}M \geq f$, with a similar change in $(\underline{1})$. Such functions are known to be BVG, Saks [1, p. 237] and we can easily prove

Theorem 10. If $f : [a, b] \to I\!R$ is measurable and if f has at least one continuous approximate bounding pair then $f \in D(a, b)$.

While this an easy generalization of Theorem 1 the corresponding generalization of Theorem 2 is false. That is: if $f \in D$ then f need not have any continuous approximate bounding pairs; this is a consequence of an example due to Tolstov [1]. The most natural generalization of the P^*-integral is Burkill's approximately continuous Perron integral, the P^*_{ap}-integral. It is defined as the P^*-integral is defined in 1.1 above except that approximate bounding pairs are used; see Burkill [1] and Bullen [1]. For this integral we do get a generalization of Theorem 2, Bullen [2].

Theorem 11. If $f : [a, b] \to I\!R$ is measurable then $f \in P^*_{\text{ap}}$ iff f has at least one approximately continuous approximate bounding pair.

It follows from Theorem 11 that if $f \in P^*_{\text{ap}}$ and has a continuous primitive then $f \in D$; Tolstov's example shows that there is an $f \in D - P^*_{\text{ap}}$.

The question of a Theorem 2 (or 11) for the D-integral remains open; the same is true for the even more general but more natural D_{ap}-integral; $f \in D_{\text{ap}}$ where there is approximately continuous ACG function such that $F'_{\text{ap}} = f$ a.e.; see Ridder [1], and Kubota [1].

In another direction Skvorcov [1] has extended Theorem 2 to the CP-integral but whether such a theorem exists for the C_nP-integral, $n > 1$, is not known.

As might be expected there are no results for any of the Perron integrals associated with various symmetric derivatives. It was pointed out by

Skvorcov that Sklyarenko, in his thesis, proved that the SCP-integral has no Marcinkiewicz theorem. This is because these integrals have only a Perron theory while the Marcinkiewicz theorem needs, as we have remarked, a descriptive or constructive theory as well.

4. BIBLIOGRAPHY

H. Bauer. 1. Der Perronsche Integralbegriff und seine Beziehung zum Lebesgueschen, Monatsh. Math. Phys., 26(1915), 153-198.

A. Bruckner. 1. Differentiation of Real Functions, Lecture Notes in Math., #659 Berlin-Heidelberg-New York, 1978.

P.S. Bullen. 1. The Burkill approximately continuous integral, J. Austral. Math. Soc., (Ser. A)35 (1983), 236-253.

2. The Burkill approximately continuous integral II, Math. Chron., 12(1983), 93-98.

P.S. Bullen and R. Vyborny. 1. Some applications of a theorem of Marcinkiewicz, to appear, Canad. Math. Bull.

J.C. Burkill. 1. The approximately continuous Perron integral, Math. Z., 34 (1931), 270-278.

V.G. Celidze and A.G. Dzvarseisvili. 1. Teoriya Integrala Denjoy i Nekotorye eë Prilozeniya, Tbilis. Gos. Univ., Tbilis, 1978. Engl. transl. to appear.

Chew, T.-S. 1. On the generalized dominated convergence theorem, Bull. Austral. Math. Soc., 37(1980), 165-171.

E.A. Coddington. 1. Theory of Ordinary Differential Equations, New York, 1955.

A. Denjoy. 1. Leçons sur le Calcul de Coefficients d'une Série Trigonométrique, Vol. IV.2, Paris, 1949.

R. Henstock. 1. Theory of Integration, London, 1963.

Y. Kubota. 1. A characterization of the approximately continuous Denjoy integral, Canad. J. Math., 22(1970), 219-226.

J. Kurzweil. 1. Nichtabsolut Konvergente Integrale, Leipzig, 1980.

Lee P.-Y. and Chew T.-S. 1. On convergence theorems for non-absolute integrals, Bull. Austral. Math. Soc., 34(1986), 133-140.

Lee P.-Y. and Lu S. 1. Notes on Classical Integration Theory (VIII), Res. Rep. #324, Dept. Math., National Univ. Singapore, 1988.

E.W. McShane. 1. On Perron integration, Bull. Amer. Math. Soc., 48(1942), 718-726.

I.P. Natanson. 1. Theory of Functions of a Real Variable, Vol. I, Engl. transl. (2nd ed. rev.), New York, 1961.

O. Perron. 1. Über den Integralbegriff, Sitzber. Heidleberg Akad. Wiss. Abt. A, 16(1914), 1-16.

J. Ridder. 1. Über approximativ stetiger Denjoy-Integrale, Fund. Math., 21 (1933), 1-10.

2. Ueber Definitionen von Perron-Integralen I, II, Indag. Math., 9(1947), 227-235, 280-289.

S.Saks. 1. Theory of the Integral, 2nd ed. rev., New York, 1937.

D.N. Sarkhel. 1. A criterion for Perron integrability, Proc. Amer. Math. Soc., 7(1978), 109-112.

A.W. Schurle. 1. A function is Perron integrable if it has locally small Riemann sums, J. Austral. Math. Soc., (Ser. A)36(1986), 220-232.

V.A. Skvorcov. 1. Some properties of CP-integrals, Amer. Math. Soc. Transl., (2)54(1966), 231-254.

G. Tolstov. 1. Sur l'intégrale de Perron, Math. Sb., 5(1939), 647-660.

University of British Columbia
Vancouver
Canada

The superposition operators in the space of Henstock-Kurzweil integrable functions

Chew Tuan Seng

Let $\mathbb{R}$ be the set of all real numbers and $[a, b]$ be a compact interval in $\mathbb{R}$. Let $k(x, t)$ be a Carathéodory function from $[a, b] \times \mathbb{R}$ to $\mathbb{R}$, i.e., the function $k(x, \cdot)$ is continuous for almost all $x \in [a, b]$ and the function $k(\cdot, t)$ is measurable for every $t \in \mathbb{R}$. Define an operator as follows:

$$P(f)(x) = k(x, f(x)).$$

Let L be the space of all Lebesgue integrable functions on $[a, b]$. It is known that P is norm-continuous if $P : L \to L$ (see [4, p.22]). In this note we shall prove the similar result for the space D of Henstock-Kurzweil integrable functions.

1. Definitions and some basic facts

Definition 1. A function F is said to be absolutely continuous on $[a, b]$ if the following condition holds:

(A) for every $\varepsilon > 0$, there is $\eta > 0$ such that for every finite or infinite sequence of non-overlapping intervals $\{[a_i, b_i]\}$ satisfying

$$\sum_i |b_i - a_i| < \eta$$

we have

$$\sum_i \omega(F; [a_i, b_i]) < \varepsilon$$

where ω denotes the oscillation of F over $[a_i, b_i]$.

Let X be a subset of $[a, b]$. A function F is said to be absolutely continuous in the restricted sense on X or, in short $AC_*(X)$ if condition A holds with the endpoints of intervals, a_i and b_i, belonging to X. Further, F is said to be generalized absolutely continuous in the restricted sense on $[a, b]$ or ACG_* if $[a, b]$ is the union of a sequence of closed sets X_i such that on each X_i the function F is $AC_*(X_i)$. Finally, a real-valued function f is said to be restricted Denjoy integrable on $[a, b]$ if there exists a function F which is continuous on $[a, b]$ and ACG_* such that is derivative $F'(x) = f(x)$ almost everywhere on $[a,b]$. It is well-known that the restricted Denjoy and Henstock-Kurzweil integrals are equivalent.

Theorem 1. (See [2, p.48, 3]) Let X be a closed subset of $[a,b]$. Then a sequence $\{f_n\}$ of functions has primitives F_n which are $AC_*(X)$ uniformly in n if

(a) for every $\varepsilon > 0$, there exists $\delta > 0$ such that

$$\left|\int_E f_n(x)dx\right| < \varepsilon \text{ for every } n$$

whenever $E \subset X$ and $m(E) < \delta$, where m is the Lebesgue measure.

(b) $\sum_{k=1}^{\infty} \omega(F_n; [a_k, b_k])$ converges uniformly in n where $\{(a_k, b_k)\}_{k=1}^{\infty}$ is the sequence of intervals contiguous to X with respect to (a,b).

Definition 2. (see [5]) A sequence $\{f_n\}$ in the space D is said to be control-convergent to f on $[a,b]$ if it satisfies

(i) $f_n(x) \to f(x)$ almost everywhere in $[a,b]$ as $n \to \infty$;

(ii) the primitives F_n of f_n are ACG_* uniformly in n, i.e., $[a,b]$ is the union of a sequence of closed sets X_i such that on each X_i the functions F_n are $AC_*(X_i)$ uniformly in n; in other words, $\eta > 0$ in the definition of $AC_*(X_i)$ is independent of n;

(iii) the primitive F_n converges uniformly on $[a,b]$.

For brevity, we shall write in what follows $F(u,v) = F(v) - F(u)$.

Theorem 2. (Controlled Convergence Theorem) (See [5]). If $\{f_n\}$ is control-convergent to f on $[a,b]$, then $f \in D$ and we have

$$\int_a^b f_n(x)dx \to \int_a^b f(x)dx \text{ as } n \to \infty.$$

The above theorem still holds if condition (iii) is replaced by the following condition (see [6]) : the primitives F_n are equi-continuous on $[a,b]$. Furthermore, condition (iii) can also be replaced by : $F_n(x)$ converges pointwise to a continuous function $F(x)$ as $n \to \infty$ (see [7]).

2. The superposition operators

Definition 3. An operator P defined on D is said to be control-continuous if

$$\int_a^b k(x, f_n(x))dx \to \int_a^b k(x, f(x))dx \text{ as } n \to \infty$$

whenever f_n is control-convergent to f in D.

Theorem 3. If $P : D \to D$, then P is control-continuous.

In proving this theorem, we may assume that $k(x,0) = 0$ for all x. Note that if

$$k_1(x,t) = k(x,t) - k(x,0)$$

and

$$P_1(f)(x) = k_1(x, f(x))$$

then

(a) $P: D \to D$ implies $P_1 : D \to D$, and

(b) P_1 is control-continuous iff P is control-continuous.

We need the following lemmas to prove the above theorem. In the following, X denotes a closed subset of $[a,b]$ and $\{(a_i, b_i)\}$ is the sequence of all intervals contiguous to X with respect to (a,b).

Lemma 1. Let $f \in D$ with $f(x) = 0$ for all $x \in X$ and $\int_{a_i}^{b_i} f \geq 0$ for all i. Then $\int_a^b F \geq 0$.

Proof. Let F be the primitive of f. Define $G(x) = F(x)$ if $x \in X$ and linearly on $(a,b) - X$. Then obviously G is ACG$_*$ and continuous. Hence $G'(x)$ exists a.e. and $G' \in D$. Furthermore $G'(x) \geq 0$ a.e., thus $\int_a^b G'(x)dx \geq 0$. Hence $\int_a^b f(x)dx \geq 0$.

Lemma 2. Let $P: D \to D$. If the primitive F of a function $f \in D$ is AC$_*(X)$, then the primitive $F^{(k)}$ of $k(x, f(x))$ is also AC$_*$(X).

Proof. Note that $P: D \to D$ implies $P: L \to D$. Hence $P: L \to L$. Therefore P transforms each set of uniformly absolutely continuous functions to a set of uniformly absolutely continuous functions (see [4, p.27]). With this property and Theorem 1, it suffices to prove that

$$\sum_i \omega(F^{(k)}; [a_i, b_i]) < \infty.$$

First, let $\{[c_i, d_i]\}$ be the sequence of subintervals of $[a_i, b_i]$ such that

$$\omega(F^{(k)}; [a_i, b_i]) = F^{(k)}(d_i) - F^{(k)}(c_i) \geq 0.$$

Let $g(x) = f(x)$ if $x \in [c_i, d_i]$ for $i = 1, 2, \ldots$ and $g(x) = 0$ elsewhere. Then $g \in D$, in view of $\sum_i \omega(F; [a_i, b_i]) < \infty$. Therefore $k(x, g(x)) \in D$. Next we shall show that

$$\sum_i [F^{(k)}(d_i) - F^{(k)}(c_i)] < \infty.$$

Let $g_n(x) = f(x)$ if $x \in [c_i, d_i]$, $i = n+1, \ldots$ and 0 elsewhere, then

$$\int_a^b k(x, g(x))dx = \sum_{i=1}^{n} \int_{c_i}^{d_i} k(x, f(x))dx + \int_a^b k(x, g_n(x))dx.$$

By Lemma 1,

$$\int_a^b k(x, g_n(x))dx \geq 0 \text{ for all } n.$$

Thus

$$\int_a^b k(x, g(x))dx \geq \sum_{i=1}^{n} \int_{c_i}^{d_i} k(x, f(x))dx, \text{ for all } n.$$

Hence

$$\sum_{i=1}^{\infty} \int_{c_i}^{d_i} k(x, f(x))dx = \sum_i [F^{(k)}(d_i) - F^{(k)}(c_i)] < \infty.$$

Similarly, we can show that

$$\sum_i [F^{(k)}(u_i) - F^{(k)}(v_i)] > -\infty$$

where

$$\omega(F^{(k)}; [a_i, b_i]) = -[F^{(k)}(u_i) - F^{(k)}(v_i)].$$

Therefore

$$\sum_i \omega(F^{(k)}; [a_i, bi]) < \infty.$$

Lemma 3. Let $P : D \to D$. If the primitives F_n of functions $f_n \in D$ are $\mathrm{AC}_*(X)$ uniformly in n, then the primitives $F_n^{(k)}$ of $k(x, f_n(x))$ are also $\mathrm{AC}_*(X)$ uniformly in n.

Proof. As in the proof of Lemma 2, we shall only show that

$$\sum_{i=1}^{\infty} \omega(F_n^{(k)}; [a_i, b_i])$$

converges uniformly in n. First, note that

$$\sum_{l=1}^{\infty} \omega(F_n; [a_l, b_l])$$

converges uniformly in n. Hence for every j, there exists $m(j)$ such that

$$\sum_{l=m(j)}^{\infty} \omega(F_n; [a_l, b_l]) \leq \frac{1}{2^j} \text{ for every } n.$$

Suppose that the theorem does not hold, i.e., there exist $\varepsilon > 0$, $p(j) \geq m(j)$ and $n(j)$ such that

$$\sum_{l=p(j)}^{\infty} \omega(F_{n(j)}^{(k)}; [a_l, b_l]) > \varepsilon.$$

Hence, for each j, there exists $q(j)$ such that

$$\sum_{l=p(j)}^{q(j)} \omega(F_{n(j)}^{(k)}; [a_l, b_l]) > \varepsilon.$$

Let $\Omega_j = \cup_{l=p(j)}^{q(j)} [a_l, b_l]$. Without loss of generality, we may assume that Ω_j are pairwise disjoint. Let $g(x) = f_{n(j)}(x)$ if $x \in \Omega_j$, $j = 1, 2, \ldots$ and 0 elsewhere. Then $g \in D$, in view of

$$\sum_{j=1}^{\infty} \sum_{l=p(j)}^{\infty} \omega(F_{n(j)}; [a_l, b_l]) < \infty.$$

Hence $k(x, g(x)) \in D$ and, by Lemma 2,

$$\sum_{j=1}^{\infty} \sum_{l=p(j)}^{q(j)} \omega(F_{n(j)}^{(k)}; [a_l, b_l]) < \infty$$

which leads to a contradiction.

Lemma 4. If $P : D \to D$, then $F_n^{(k)}$ are equicontinuous whenever F_n are equicontinuous.

Proof. Suppose that F_n are equicontinuous. Then for every $x \in [a, b]$, there exists $\delta_j > 0$ such that (A)

$$|F_n(y) - F_n(x)| \leq \frac{1}{2^j} \text{ for every } n$$

whenever $|y - x| \leq \delta_j$. We shall prove that $F_n^{(k)}$ are equicontinuous. Suppose that it is not true, then there exist $x \in [a, b]$ and $\varepsilon > 0$; for every $\delta_j > 0$, there exist $y(j) > 0$ with $|x - y(j)| \leq \delta_j$ and $n(j)$ such that

$$|F_{n(j)}^{(k)}(x) - F_{n(j)}^{(k)}(y(j))| > \varepsilon.$$

Note that for any fixed $n(j)$, $F_{n(j)}^{(k)}$ is continuous. Hence there exists $z(j)$ with $0 < |x - z(j)| \leq \delta_j$ such that

$$|F_{n(j)}^{(k)}(y(j)) - F_{n(j)}^{(k)}(z(j))| > \varepsilon.$$

We may choose δ_j such that the sequence of closed intervals I_j with end-points $y(j)$ and $z(j)$ are pairwise disjoint. In view of (A)

$$\sum_j \omega(F_{n(j)}; I_j) < \infty.$$

By Lemma 2

$$\sum_j \omega(F_{n(j)}^{(k)}, I_j) < \infty$$

which leads to a contradiction.

Proof of Theorem 3.

Let $\{f_n\}$ be a sequence of functions in D which is control-convergent to f, then, by the above lemmas, $\{k(x, f_n(x))\}$ is control-convergent to $k(x, f(x))$. Hence, by the controlled convergence theorem, P is control-continuous.

3. Representation theorem

Definition 4. A functional G defined on D is said to be orthogonally additive if $G(f + g) = G(f) + G(g)$ whenever f and g have disjoint suppports.

Theorem 4. A functional G is orthogonally additive and control-continuous on D iff there exists a Carathéodory function $k(x,t)$ with $k(x,0) = 0$ a.e. such that $k(x, f(x)) \in D$, for every $f \in D$ and

$$G(f) = \int_a^b k(x, f(x))dx \text{ for all } f \in D.$$

Proof. The necessity is proved in [3, Theorem 5], however the proof of the necessity can be simplified by using the controlled convergence theorem in [7, Theorem 2]. The sufficiency follows from Theorem 3.

The above representation theorem is much better than that in [3, Theorem 5]. The linear case (see [1], [8], [3]) can be deduced immediately without resorting to the Sargent's technique [8,3].

References

[1] A. Alexiewicz, Linear functionals on Denjoy integrable functions, Coll. Math. 1(1948) 289-293

[2] V. G. Chelidze and A. G. Dzhvarsheishvili, Theory of the Denjoy integral and some of its applications (Russian), Tbilisi, 1978.

[3] T.S. Chew, Nonlinear Henstock integrals and representation theorems, SEA Bull. Math. 12(1988), 97-108.

[4] M. A. Kransnosel'skii, Topological methods in the theory of nonlinear integral equations (translation), Pergamon Press, New York, 1964.

[5] P. Y. Lee and T. S. Chew, A short proof of the controlled convergence theorem for Henstock integrals, Bull. London Math. Soc. 19(1987) 60-62.

[6] P. Y. Lee and T. S. Chew. a Riesz-type definition of the Denjoy integral, Real Anal. Exchange 11(1985/86) No. 1, 221-227.

[7] K. C. Liao, A refinement of the controlled convergence theorem for Henstock integrals, SEA Bull. Math. 11(1987) 49-51.

[8] W. L. C. Sargent, On the integrability of a product, J. London Math. Soc. 23(1948) 28-34.

National University of Singapore
Singapore

New and old results concerning Henstock's integrals

Susana Fernandez Long de Foglio

1. SURVEY OF A SAMPLE OF DEFINITIONS OF RIEMANN TYPE INTEGRALS

1.1. After Lebesgue gave his definition of integral many attempts were made to define an integral equivalent to it but using a process of sums of Riemann type in a natural way. Lebesgue had proved that: *"If f is a summable function on $[a, b]$ then there exists a net consisting of the sequence of partitions $\{\sigma_n\}$, $\sigma_n = \{x_i^n\}_0^{i_n}$ with diameter tending to zero, and of the points $\{\xi_i^n\}_i^{i_n}$ such that the Riemann sums $\sum_{i=1}^{i_n} f(\xi_i^n)(x_i^n - x_{i-1}^n)$ converge to the Lebesgue integral $\int_a^b f(x)dx$ as $n \to \infty$"* [19 page 103]. (The assertion cannot be considered as a definition.)

In the references some of the many definitions which appear from 1919 can be seen. Some of these generalize the Lebesgue definition. They can be grouped in two classes:

- All the partitions are allowed and the integral is defined using major and minor functions whose values are calculated by excluding some values which the functions take on some sets.
- Not all partitions are allowed.

The sample of definitions chosen in this survey are those of:

1. B. Levi [16]
2. E.J. McShane [1]
3. A. Denjoy [2] definition (A) and [3]
4. The Riemann Complete Integral [10,13]

The sample chosen represents the "feeling" of the search. With the inclusion of B. Levi's definition which is almost unknown – it was published in Spanish – I hope to give some interest to this survey.

1.2 The Beppo Levi Integral

The definition given by B. Levi is valid for functions in $\mathbb{R}^n$ but, without detriment to our purpose, it is sufficient to give it for functions $f : [a, b] \to \mathbb{R}$.

Let $[a, b]$ be an interval in $I\!R$ and $f: [a, b] \to I\!R^+$. The following definitions are needed.

1. A finite or enumerable family of non-degenerate intervals Σ of $I\!R$ is a ***covering*** for $[a, b]$ if each point in $[a, b]$ is interior to some of these intervals. (Without loss of generality, we can take the intervals to be $[a, x)$ and $(y, b]$; $a < x < b$, $a < y < b$.)

2. If to each interval I of the system Σ is associated a number $h(I) > 0$ we have defined a system of ***rectangles*** $I \times [0, h(I)]$.

3. Given a covering Σ, to each point $x \in [a\,b]$ there corresponds an interval, i.e., there exists a function $\Phi: [a, b] \to \Sigma$ and $\Phi(x) = I$ for some $I \in \Sigma$.

4. A system of ***prevalent rectangles*** for a function f is any system of rectangles such that their bases form a covering Σ for $[a, b]$ and $f(x) \leq h(I) = h(\Phi(x))$. The area of such a system is called a ***superior sum*** of $f(x)$ in $[a, b]$.

Definition 1.2.1. Let f be a non-negative bounded function $0 \leq f(x) \leq M < \infty \quad \forall x \in [a, b]$.

1. The superior integral of f in $[a, b]$ is the greatest lower bound of the superior sums on $[a, b]$ and is denoted by
$$S = \overline{\int_a^b} f(x)dx.$$

2. If $f(x)$ takes some negative values and $|f| < M$ then the superior integral is defined by
$$\overline{\int_a^b} f(x)dx = \overline{\int_a^b} (f(x) + M) - M(b - a).$$

3. The inferior integral of $f(x)$ in $[a, b]$, $|f(x)| \leq M$ is defined by
$$s = \underline{\int_a^b} f(x)dx = -\overline{\int_a^b} (-f(x))dx.$$

4. If $\overline{\int_a^b} = \underline{\int_a^b}$ then $f(x)$ is said to be integrable and their common value is the integral (in the B.L. sense) of $f(x)$ over $[a, b]$.

The inferior integral cannot be given via inferior sums with $h(I) \leq f(x)$ because for different coverings it may happen that an inferior sum is bigger than a superior sum, as the following example shows.

Let $f(x)$ be defined on $[0, 1]$ by

$$\begin{aligned} f(x) &= 0 \text{ if } x \in [0, 1],\ x \neq 1/2, \\ &= -1 \text{ if } x = 1/2. \end{aligned}$$

Let Σ, Φ be as follows.

$$\begin{aligned} \Sigma &= \{[0, 2/3], [1/3, 1], [0, 1]\}, \\ \Phi(x) &= [0, 2/3]\ \forall x \in (0, 2/3),\ x \neq 1/2, \\ \Phi(x) &= [1/3, 1] \text{ if } x \in [2/3, 1], \\ \Phi(1/2) &= [0, 1], \end{aligned}$$

For Σ, Φ the superior sum is -1. If, on the other hand, we force Σ to have diameter less than 1, we can easily produce an inferior sum greater than -1.

B. Levi points out [16 page 94] that to have all the desired properties for the superior integral it is necessary to use all the possible prevalent rectangles and not only some special families. When b is unbounded the integral $\int_a^\infty f(x)dx$ is defined as $\lim \int_a^A f(x)dx$ as $A \to \infty$. For the general case, $f(x)$ unbounded, B. Levi suggests in [16] various possibilities.

1.3. McShane's definition of the Itô-belated deterministic integral

The divisions used by McShane are in certain sense opposed to those used by B.Levi. McShane covers "too little" instead of the latter who covers "too much".

Definition 1.3.1 Let $f : [a, b) \to \mathbb{R}$, $\gamma : [a, b) \to \mathbb{R}^+$ ($\gamma > 0$ almost everywhere), and $\delta > 0$. Then $\pi = \{(A_i, x_i) : i \leq m\}$ is a (γ, δ) partition of $[a, b)$ iff

1. $A_i = [a_i, b_i) \subset [a, b)$ $\forall i$ such that $1 \leq i \leq n$,

2. $A_i \cap A_j = \emptyset$ when $i \neq j$,

3. $A_i \subset [x_i, x_i + \gamma(x_i))$, $1 \leq i \leq n$,

4. $m([a, b) - \cup_{i-1}^n A_i) < \delta$,

5. $\Sigma(f, \pi) = \sum_{i=1}^n f(x_i) m(A_i)$

The function f is integrable in the deterministic ***Itô-belated*** sense iff $\forall \varepsilon > 0$ $\exists \gamma, \delta$ such that if π_1 and π_2 are (γ, δ)-partitions in $[a, b)$ then $|\Sigma(f; \pi_1) - \Sigma(f; \pi_2)| < \varepsilon$. The pair (γ, δ) is said to correspond to ε for f. The deterministic Itô-belated integral is equivalent to the Lebesgue integral [1, 5].

Its principal features are

- Not all partitions are allowed.

- The partitions do not necessarily cover the domain of integration.

1.4. The Denjoy (A)-integral

In 1919 [2] A. Denjoy gave three definitions of Riemann-type integrals, called by him (A), (B) and (C) integrals, which contained the Lebesgue definition of integral. In 1931 [3] A. Denjoy proved that the (A)-integral was equivalent to the Lebesgue integral (see also [19] p.104).

The definition of the latter suggests an important property of the Lebesgue integral, used in calculation of integrals, namely the Monte-Carlo method. Denjoy uses the notion of measure. The definition is as follows. Let $f(x)$ be defined on a bounded interval $[a, b] \subset \mathbb{R}$ and let $a, x_1, x_2, \ldots, x_{n-1}, b$ be any partition of the interval $[a, b]$. The superior and inferior sums $\sum M_i(x_i - x_{i-1})$, $\sum m_i(x_i, x_{i-1})$ are obtained by the following steps.

1. Fix a pair of numbers (α, β), $\alpha > 0, \beta > 0, \alpha + \beta < 1$.

2. The mean density of a set E in $[a, b]$ is defined as $\mu(E)/(b-a)$, μ being the measure of E, and, in a analogous way, $\mu(E \cap (x_{i-1}, x_i))/(x_i - x_{i-1})$ is the mean density of E in (x_{i-1}, x_i).

3. M is the maximum of mean density α in $[a, b]$ if M is the greatest number y such that
$$\frac{\mu E_x(f \geq y)}{b-a} \leq \alpha$$
and m is the minimum of mean density β in $[a, b]$ if m is the smallest number y such that
$$\frac{\mu E_x(f \leq y)}{b-a} \leq \beta.$$
Similarly for each sub-interval (x_{i-1}, x_i).

4. f is said to be (A) integrable on $[a, b]$ if the superior sums and the inferior sums have the same limit when the maximun of $x_i - x_{i-1}$ goes to zero.

The main features of Denjoy's (A)-integral are:

- Any division is allowed.
- It uses major and minor sums.
- For the computation of the major and minor sums only the values of f on certain sets are used.

The following important property of the Lebesgue integral (uniform convergence of samples) is easily derived from the definition of the (A)-integral:

If $f : [a, b] \to \mathbb{R}^n$ is summable with $\int_a^b f d\mu = I$ where μ is the Lebesgue measure, then, for $\lambda \in [0, 1]$,

$$g_k(\lambda) = \sum_{n=1}^{2^k} \int_{(n-1)/2^k}^{(n-1+\lambda)/2^k} f d\mu$$

converges to λI uniformly in λ as k tends to infinity [9].

1.5. Definition of the Henstock-Kurzweil integral

R. Henstock [10] and J. Kurzweil [13] have defined an integral using finite partitions of Riemann type. All values of the function are considered. The definition shares with the definition we gave in 1.3, the property that functions $\delta : [a, b] \to \mathbb{R}^+$ define the partitions. Theorem 16.1 in [10] ensures the existence of divisions of $[a, b]$. The definition is simpler than the four given above. Let $f : [a, b] \to \mathbb{R}$. The function f is said to be integrable on $[a, b]$ if there exists a real number ω such that $\forall \varepsilon > 0$ there exists a function $\delta : [a, b] \to (0, \infty)$ such that

$$|\omega - \sum_{i=1}^{k} f(t_i)(x_i - x_{i-1})| < \varepsilon$$

provided

$$a = x_0 \leq t_i \leq x_2 \leq \ldots \leq x_{k-1} \leq t_k \leq x_k = b,$$

$$[x_{i-1}, x_i] \subset (t_i - \delta(t_i), t_i + \delta(t_i)).$$

The number ω is the integral of f on $[a, b]$. This integral was called by R. Henstock the Riemann-complete integral and denoted by (RC) $\int_a^b f(x)dx = \omega$. The Riemann-complete integral is more general than the Lebesgue integral. It is equivalent to the Perron and special Denjoy integrals. It can be defined for several variables and it has been extended to abstract spaces. We suggest that the problem in 1.1 is now completely solved. The (RC)-integral uses Riemann sums in a simple and natural way. The fact that it is more general than the Lebesgue integral is an advantage which is used already in the field of differential equations [18].

2. THE N-VARIATIONAL INTEGRAL AND THE SCHWARTZ DISTRIBUTION AND COLOMBEAU GENERALIZED FUNCTIONS

2.1 In [11] R. Henstock defined several integrals of the Stieltjes type, proving their equivalence. One of them is the N-variational integral. To define the N-variational integral a finite iteration is used. We start with Ward integration $W = N_0$. Then we show how N_i is got from $N_{i-1}, = S$. First, we need the following definition. A function F, defined and finite in $[a, b]$, is a N-major function of f, ϕ in $[a, b]$ if $F(a) = 0$ and if

$$(S)\int_{-h}^{0}\{F(x)-F(x+t)\}d_tN(x,-h;t) \geq f(x)(S)\int_{-h}^{0}\{\phi(x)-\phi(x+t)\}d_tN(x,-h;t)$$

for all intervals $(x - h, x)$ in a left complete family in $[a, b]$, with an analogous condition to the right. Let

$$N(x, h; t) \;\; (0 \leq t \leq h, \; a \leq x \leq x + h \leq b)$$

and

$$N(x, -h; t) \ (-h \leq t \leq 0, \ a \leq x - h \leq x \leq b)$$

be such that, for $a \leq x + h \leq b$,

$$N(x, h; h) = 1, \ N(x, h; 0) = N(x, h; 0+) = 0; \tag{1}$$

for $a \leq x - h \leq b$,

$$N(x, -h; 0) = N(x, h; 0-) = 1, \ N(x, -h, -h) = 0; \tag{2}$$

$$N(x, h; t) \text{ and } N(x, -h; t) \text{ are monotone increasing in } t; \tag{3}$$

if F is an N-major function of 0, ψ in $[a, b]$ then

$$F \text{ is monotone increasing there.} \tag{4}$$

The intervals $(x, x+h)$ and $(x-h, x)$ are defined as for the RC-integration [10, 11], i.e. a family $\mathcal{L}$ of intervals is left complete in $[a, b]$ if there is an $h_1(\mathcal{L}, x) = h_1(x) > 0$ and $a < x \leq b$ such that every interval $(x-h, x)$ in $[a, b]$ with $0 < h \leq h_1(x)$ lies in $\mathcal{L}$ with an analogous definition for $\mathcal{R}$, $h_2(\mathcal{R}, x)$; $\mathcal{R}$ the right complete family in $[a, b]$.

Without loss of generality we can take $h_1(x) = h_2(x) = h(x)$. A pair $k = (k_l, k_r)$ of interval function is of bounded N-variation over $[a, b]$ if there exist a function $\delta : [a, b) \to (0, \infty)$ which defines families of intervals to the right and to the left of each point, and a monotone increasing function χ, such that

$$|(\mathrm{S})\int_0^h k_r(x, x+t) d_t N(x, h; t)| \leq (\mathrm{S})\int_0^h \{\chi(x+t) - \chi(x)\} d_t N(x, h; t) \tag{5}$$

$$((x, x+h) \in \mathcal{R} \text{ i.e., } h \leq \delta(x))$$

and equivalent relations for intervals $(x-h, x) \in \mathcal{L}$, [6].

The N-variation of $\mathbf{k}$ over $[a, b]$ is defined to be

$$V(N; \mathbf{k}; [a, b]) = \inf\{\chi(b) - \chi(a)\}$$

for all such χ. Further, $\mathbf{k}$ is N-variationally equivalent to 0 in $[a, b]$, if $V(N; \mathbf{k}; [a, b]) = 0$. Then, given $\varepsilon > 0$ we can choose the χ in (5) to satisfy $0 \leq \chi(b) - \chi(a) < \varepsilon$. Given three functions H, ψ, f in that order, we say that H is N-variationally equivalent to ψ, f in $[a, b]$, if $\mathbf{k}$ is N- variationally equivalent to 0, where

$$k_l(u, v) = H(v) - H(u) - \psi(v)(f(v) - f(u))$$

$$k_r(u, v) = H(v) - H(u) - \psi(u)(f(v) - f(u))$$

The difference $H(b) - H(a)$ is called the N-variational integral of ψ, f in $[a, b]$.

In [6,7,8] some Schwartz distributions were represented by N-variational integrals ((NV)-integrals), in particular those with discrete and compact support.

In [6] the representation of the generalized function δ' by a (NV)-integral was given by

$$< \delta', \psi >= (\text{NV}) \int_{-\infty}^{\infty} \psi(x) df(x) \tag{6}$$

where $\psi \in \mathcal{D}$ and $f(x) = x^{-1}\text{ch}(A)$, $A = \{x_n\}$ with $x_n \to 0$ as $n \to \infty$ and the family $N(x, h; t)$ was given,if $x \neq 0$, by

$$\begin{aligned} N(x,\ h;\ t) &= t/h \\ N(x,\ -h;\ t) &= 1 + t/h \end{aligned} \tag{7}$$

To define $N(0, h; t)$ we use the set A. If $x(h)$ is the greatest x_n in the range $0 < x_n < h$ we put

$$\begin{aligned} N(0,\ h;\ t) &= 0\ (t < x(h)), \\ &= 1\ (t \geq x(h)). \end{aligned} \tag{8}$$

Since the sequence $\{x_n\}$ is arbitrary the definition (6) is not unique. Instead of a sequence, the set A can be taken as any set of null-measure (taking some care in (8)). So a class of functions f and families N define such a distribution. Since δ' is given as a Stieltjes-type integral we can associate $f(x)$ to the distribution δ. In this way $f(x)$ gives the "behaviour" of δ at zero. We can think of $\delta(0)$ as an infinitely large value defined by $x^{-1}\text{ch}A$. We will this in part 3.

2.2 J.F. Colombeau [4] has extended the concept of distribution by defining the generalized functions $\mathcal{G}$ in which a solution was found of the problem of multiplication of the Schwartz distributions. $\mathcal{G}$ turns out to be an associative and commutative algebra with partial differentiation, integration and other properties. T.V. Todorov [20,21], preserving all "good properties of the Colombeau functions" and with a slight modification of the theory, has given a nonstandard version of it.

In this section we calculate an (NV)-integral which will be associated with the generalized function $(\delta')^2$ of the Colombeau theory. The rationale of this association will be explained in part 3. In what follows the principal steps in the construction of such an (NV)-integral are given. R. Henstock has defined the generalized Roussel derivative [11] to the right (left) of a function H with respect to a function f at one point using g (g is an interval function with certain properties [11]) as a number k such that

$$(\text{S}) \int_0^h [H - kf]_x^{x+t} d_t N(x,\ h;\ t)/g(x,\ x+h) \to 0 \text{ as } h \to 0. \tag{9}$$

Theorem (Th. 8, [11]): If $I(x) = I(x;\ \psi,\ f;\ a_1 x)$ $(a \leq x \leq b)$ (i.e., $I(x) = (\text{NV}) \int_a^x \psi(x) df(x)$) exists, $D(N;\ g;\ I,\ f;\ x) = \psi(x)$, except possibly at points of a set E, such that $\mathcal{A}_1(E) = 0$ (where $\mathcal{A}_1(A)$ is a Carathéodory outer measure related to g [11]).

2.3 Construction of $(\delta'(\psi))^2$.

From [6] we have that $<\delta', \psi> = (\mathrm{NV}) \int_{-\infty}^{\infty} \psi(x)df(x)$, ψ, f as given in (6), $N(x, \pm h; t)$ as given in (7,8). Let $F(x)$ be the function

$$F(x) = (\mathrm{NV}) \int_{-\infty}^{x} \psi(x)df(x), \ \psi \in \mathcal{D}, \ \psi \text{ fixed.} \tag{10}$$

$$\begin{aligned} F(x) &= -\psi'(0) \text{ if } x > 0, \ x \notin A = \{x_n\}, \\ &= -\psi'(0) + \frac{\psi(x_n)}{x_n} \text{ if } x_n \in A, \\ &= 0 \text{ if } x \leq 0; \\ F^2(x) &= (\psi'(0)^2 \text{ if } x > 0, \ x \notin A, \\ &= (-\psi'(0) + \frac{\psi(x_n)}{x_n}^2 \text{ if } x_n \in A, \\ &= 0 \text{ if } x \leq 0. \end{aligned} \tag{11}$$

Let $F^2(x) = \int_0^x \phi(x)df^2$ be such that $\phi(x) = \frac{dF^2}{d(f^2)}$ if it exists in the generalized Roussel sense. We will see that such a $\phi(x)$ cannot exist, but with a minor modification to $\phi(x)$ we will be able to express

$$(\psi'(0))^2 = (\mathrm{NV}) \int_{-\infty}^{\infty} \overline{\phi}(x)df^2;$$

$$\overline{\phi}(x) = \phi(x) + \phi_\epsilon(x).$$

For $x = 0$, $\phi(0)$ must be "infinite" and then the generalized Roussel derivative cannot exist. Applying (9) to $F^2(x)$ we have: If $x > 0$ and $x \notin A = \{x_n\}$ then $[F^2(x+t) - F(x)^2 - kf]_x^{x+h} = 0$, $x + t \notin A$ and the Roussel derivative is 0 for any g_1 possible. If $x \in A$ then the increment

$$\begin{aligned} [F(x+t) - F(x) - kf]_x^{x+t} &= (-\psi'(0))^2(-\psi'(0) + \psi(x)/x)^2 + k(1/x^2) \\ &= -\psi(x)^2/x^2 + 2(\psi'(0)\psi(x).x/x^2 - k(1/x^2) \\ &= \Delta = 0 \end{aligned}$$

for $k = \psi^2(x) - 2\psi'(0)\psi(x).x$ and for any g_1,

$$(\mathrm{S}) \int_0^h \Delta \, d_t N(x, h; t)/g_1(x, x+h)$$

tends to 0 as $h \to 0$. After verifying that the NV-integrals below exist, and applying theorem 8 [11], we have

$$\begin{aligned} (\mathrm{NV}) \int_{x_n}^{\infty} \phi df^2 &= (\mathrm{NV}) \int_{x_n}^{\infty} \frac{dF^2}{df^2} df^2 \\ &= F^2(\infty) - F^2(x_n) \\ &= 2\psi'(0)\frac{\psi(x_n)}{x_n} - \frac{\psi^2(x_n)}{x_n^2}. \end{aligned}$$

We are interested to define a function $\overline{\phi}$ such that

$$(\mathrm{NV})\int_{-\infty}^{\infty}\overline{\phi}(x)df^2=(\psi'(0))^2.$$

For this purpose we make the following calculations [11].

$$(\mathrm{S})\int_0^h(\int_{x_n}^{\infty}\phi df^2+\phi(0)(f^2(x_n)-f^2(0))d_tN$$

$$=2\psi'(0)\tfrac{\psi(x_n)}{x_n}-(\tfrac{\psi(x_n)}{x_n})^2+\tfrac{\psi(0)^2}{x_n^2}$$

$$=2\psi'(0)\tfrac{\psi(x_n)}{x_n}+(\tfrac{\psi(x_n)+\psi(0)}{x_n})(\tfrac{\psi(0)-\psi(x_n)}{x_n})$$

$$=\psi'(0)(2\tfrac{\psi(x_n)}{x_n})+(\tfrac{\psi(x_n)+\psi(0)}{x_n})(-\psi'(0)+\varepsilon)$$

$$=\psi'(0)(2\tfrac{\psi(x_n)}{x_n})-(\tfrac{\psi(x_n)+\psi(0)}{x_n})+\varepsilon\tfrac{\psi(x_n)+\psi(0)}{x_n}$$

$$=\psi'(0)(\tfrac{2\psi(x)-2\psi(x)}{x_n}+\tfrac{\psi(x_n)-\psi(0)}{x_n})+\varepsilon\tfrac{\psi(x_n)+\psi(0)}{x_n}$$

$$=\psi'(0)(\psi'(0)-\varepsilon)+\varepsilon\tfrac{(\psi(x_n)+\psi(0)}{x_n}.$$

The term $\varepsilon(x)\frac{\psi(x)+\psi(0)}{x}$; $x\in A=\{x_n\}$ where

$$\varepsilon(x)=\frac{\psi(x)-\psi(0)}{x}-\psi'(0)$$

has not in general a finite limit as $x_n\to 0$. To obtain the desired result we need to modify ϕ by adding an extra term, i.e., $\overline{\phi}=\phi+\phi_\varepsilon$ where

$$\begin{aligned}\phi_\varepsilon(x)&=-2\varepsilon(x_n)\psi(x_n)x_n \text{ if } x_n\in A\\&=0 \text{ otherwise.}\end{aligned}\qquad(12)$$

and then

$$\begin{aligned}&(\mathrm{NV})\textstyle\int_{x_n}^{\infty}(\phi+\phi_\varepsilon)df^2\\&=\psi'(0)(\psi'(0)-\varepsilon(x_n))+\tfrac{\varepsilon(x)(\psi(x_n)+\psi(0))}{x_n}+2\tfrac{\varepsilon(x_n)\psi(x_n)}{x_n}\end{aligned}$$

$$=\psi'(0)(\psi'(0)-\varepsilon(x_n))+\tfrac{\varepsilon(x_n)\psi(0)}{x_n}-\tfrac{\varepsilon(x_n)\psi(x_n)}{x_n}$$

$$\begin{aligned}&=\psi'(0)(\psi'(0)-\varepsilon(x_n))+\varepsilon(x_n)\tfrac{\psi(0)-\psi(x_n)}{x_n}\\&\to(\psi'(0)^2 \text{ as } x_n\to 0.\end{aligned}$$

Then

$$(\text{NV})\int_{-\infty}^{\infty}(\phi+\phi_\varepsilon df^2=(\psi'(0))^2. \tag{13}$$

Many other (NV) integral expressions can be obtained to represent $(\psi'(0))^2$, each yielding an equivalent kind of "infinite value" at zero.

3. THE "INFINITE" OF THE FUNCTIONS USED IN PART 2 AND THEIR NON- STANDARD REPRESENTATION

In Colombeau [4] the following definitions are given: Let $\mathcal{D}(\mathbb{R}^n)$ be the space of all C^∞ functions of $\mathbb{R}^n \to \mathbf{C}$ with compact support, and

- $\mathcal{A}_q = \{\psi \in \mathcal{D}(\mathbb{R}^n) : \int_{\mathbb{R}^n}\psi(\lambda)d\lambda = 1,\ \int_{\mathbb{R}^n}\lambda^i\psi(\lambda)d\lambda = 0,\ 1 \leq |i| \leq q\}$,
- $\psi_\varepsilon(\lambda) = \frac{1}{\varepsilon^n}\psi(\frac{\lambda}{\varepsilon})$ where $\varepsilon > 0$, $\lambda \in \mathbb{R}^n$ and $\psi \in \mathcal{D}(\mathbb{R}^n)$,
- $\mathcal{E}[\mathbb{R}^n]$ is the algebra of all functions $R : \mathcal{A}_1 \times \mathbb{R}^n \to \mathbf{C}$ which are C^∞ in x $\forall\psi$ fixed.
- A generalized function is an element of the quotient space

$$\mathcal{G}(\mathbb{R}^n) = \mathcal{E}_M\mathbb{R}^n/\mathcal{N}(\mathbb{R}^n).$$

$\mathcal{E}_M[\mathbb{R}^n]$ is a linear subspace and a subalgebra of $\mathcal{E}[\mathbb{R}^n]$ and $R \in \mathcal{E}_M[\mathbb{R}^n]$ iff $\forall\psi \in \mathcal{A}_N$ $\exists\eta > 0$, $c > 0$ such that

$$|DR(\psi_\varepsilon, x)| < \frac{c}{\varepsilon^N}, \quad (D = \frac{\partial^{|k|}}{\partial_{x_1}^{k_1}\dots\partial_{x_n}^{k_n}}).$$

The set $\mathcal{N}$ is an ideal of $\mathcal{E}[\mathbb{R}^n]$ defined by the functions of $\mathcal{E}[\mathbb{R}^n]$ such that: for each compact subset K of $\mathbb{R}^n$ and each derivation operator D, $\exists N \in \mathbb{N}$, α an increasing function from $\mathbb{N} \to \mathbb{R}^+$, $\alpha(q) \to \infty$ as $q \to \infty$, so that if $\psi \in \mathcal{A}_q$, $q \geq N$, then $\exists\eta > 0$, $c > 0$ with $|(\mathrm{DR})(\psi_\varepsilon, x)| < c\varepsilon^{\alpha(q)-N}$.

Continuous functions and all distributions are properly embedded in $\mathcal{G}$ and their multiplication is well defined in $\mathcal{G}$. The values of such $R \in \mathcal{G}$ belong to an algebra which is an extension of the complex numbers $\mathbf{C}$ with infinite and infinitesimal numbers, but the algebra is not a field.

T.D. Todorov [20, 21] using an ultrafilter with some special properties on $\mathcal{D}_+ = \mathcal{D} \times \mathbb{R}^+$, and modifying slightly Colombeau's theory, has defined meta-functions which possess all the good properties of Colombeau's new generalized functions, and an extension of the complex numbers which is a non-standard field (the moderate hypercomplex numbers).

The meta-functions are the "moderate elements" $*\mathcal{E}_M$ of $*\mathcal{E}$, which is the non-standard version of $\mathcal{E} = C^\infty(\mathbb{R}^\backslash)$. With the ideas given above we will try to analyze the behaviour of the NV- integrals defined in Part 2.

In the definition of $\psi'(0)$ ((6) and [6]) the value of $f(0)$ is zero but its influence in zero is governed by the sequence x^{-1} ch A, $A = \{x_n\}$, so the value of f at $x = 0$ can be associated with the behaviour of x^{-1} ch A. The set A could be any set of null measure with zero as one of its limit points. The values of f in $x \in A$ have no relation to the behaviour of $f(x)$ in x since such a set A is arbitrary with the conditions given above. We can think of f not as a fixed function but a family of functions, the functions used in (6) being one of its representants.

J.F. Colombeau [4] has defined an integral process for the new generalized functions and he has proved that in the particular case of the distributions:

$$\int T\psi dx = < T, \psi > .$$

The distributions and their values are associated with the nucleus T. Analogous considerations can be applied to the NV-integrals which represent distributions. Furthermore since the distributions by NV integrals are of the Stieltjes type the correspondence of the integrator must be, not with the distribution T which it represents, but with the one such that its derivative is T. In the particular case of (6) we have

$$< \delta', \psi >= (\text{NV}) \int_{-\infty}^{\infty} \psi(x) df(x)$$

with f and N defined in (7), (8).

Such $f(x)$ may be associated with δ and the value of $\delta(0)$ associated with the "infinite number" defined by x^{-1} ch A. We have also used the NV-integral to represent $(\delta')^2$ in Part 2. In this case, with Colombeau's notation,

$$R(\psi, x) = (\delta')^2 = (\psi'(0))^2.$$

Our representation uses for integrator a function f^2 which has an "infinite" value at zero defined in a way analogous to the above, i.e., as a class of functions which represent $R(\psi, x)$; but now $R(\psi, x)$ is not a distribution and ψ appears in a different way than in (6),so the meaning of our integral cannot be taken as we have done above for distributions.

The NV-integrals appear to be connected with more general objects than distributions.

REFERENCES

[1] DARTS, R.B. and McSHANE, E., *The deterministic Itô-belated integral is equivalent to the Lebesgue integral*, Proc. of the Amer. Math. Soc., vol. 72, number 2, (1978).

[2] DENJOY, A., *Sur l'integration riemanienne*, Comptes Rendus 169, 219-221 (1919).

[3] DENJOY, A., *Sur la definition riemanienne de l'integral de Lebesgue*, Comptes Rendues 193, (1931), 695-698.

[4] COLOMBEAU, J.F., *Elementary introduction to new generalized functions*, North-Holland Math. Studies, 113 (1985).

[5] FOGLIO, S., *A integral determinística Itô-belated segundo uma velha definitiva de Beppo Levi*, Atas do 5º Simpósio Nacional de Probabilidade e Estatística, (1982), (Brazil).

[6] FOGLIO, S., *The N-variational integral and the Schwartz distributions*, Proc. London Math. Soc., 18, (1968), 337-348.

[7] FOGLIO, S., *The N-variational integral and the Schwartz distributions II*, J. London Math. Soc. (2), (1970), 14-18.

[8] FOGLIO, S. and HENSTOCK, R., *The N-variational integral and the Schwartz distributions III*, J. London Math. Soc. (2), (1973), 693-700.

[9] HALKIN, H., *On the uniform limit of multiple balayage of vector integrals*, Amer. Math. Monthly, vol. 73, pp. 733-735, (1966).

[10] HENSTOCK, R., Theory of integration, Butterworth, London, (1963).

[11] HENSTOCK, R., *The equivalence of generalized forms of the Ward, Variational, Denjoy-Stieltjes, and Perron-Stieltjes integrals*, Proc. London Math. Soc. (3), 10, (1960).

[12] KEMPISTY, S., *Sur l'integrale (A) de M. Denjoy*, Compt. Rend., 185, (1927), 749-751.

[13] KURZWEIL, J., *Generalized ordinary differential equations and continuous dependence on a parameter*, Czech. Math., J., 7 (82), 1957, 568-583.

[14] LEVI, B., *Sulla definizione dell'integrale*, Annali di matematica, Serie IV, T.I., 1924.

[15] LEVI, B., *Sulla definizione della integrale*, Scritti Matematica, Oferti a Luiggi Berzolari, Pascua, 1936.

[16] LEVI, B., *Teoria de la integral de Lebesgue, independiente de la noción de media*, Publicaciones del Inst. de la Univ. del Litoral (Argentina), (1941).

[17] McSHANE, E.J., Stochastic calculus and Stochastic models, Acad. Press, 1974.

[18] MULDOWNEY, P., A general theory of integration in function spaces, Longman Scient. Tech. Pitman Research Notes in Math. Series, (1987).

[19] PESIN, J.N., Classical and Modern Integration theories, Acad. Press, 1970.

[20] TODOROV, T.D., Sequential approach to Colombeau's theory of generalized functions, Miramare, Trieste, (1987).

[21] TODOROV, T.D., Colombeau's generalized functions and non-standard analysis, Miramare, Trieste, (1987).

UNIVERSIDADE FEDERAL DE DE SÃO CARLOS
DEPARTMENTO DE ESTATÍSTICA
SÃO CARLOS, SP, BRAZIL

Double integrals and convergence of double series

Cecile Pierson-Gorez

0. INTRODUCTION

The connection between the Perron integral and the convergence of simple series is already well known and its proof can be found in, for example, [MAW1]. A relation between the Lebesgue integral and convergence of double series with positive coefficients has been also discovered. In this paper, we define three new double integrals which allow us to establish a connection between integrability and convergence of arbitrary double series. These integrals are based on an integral discovered recently by Buczolich [BU] and Mkhalfi [MKH] and called the (GM)-integral.

The definition of the (GM)-integral and of other concepts is given in Section 1. Section 2 contains the definition of the new integrals over unbounded intervals and their basic properties, especially a generalisation in two dimensions of Hake's theorem. The last section consists of a proof of the connection mentioned above.

1. IRREGULARITY AND (GM)-INTEGRAL

First, we denote by $d(A)$, $\text{int}(A)$, $\text{Cl}(A)$ the diameter, interior and closure of a set A and we work with norm $|x| = \max(|x_1|, |x_2|)$ in $\mathbb{R}^2$. Let $I = [a_1, b_1[\times[a_2, b_2[$ be a left closed bounded interval of $\mathbb{R}^2$. As usual, we define a P- partition and a δ-fine P-partition of I.

Definition 1: A P-partition of I is a finite family

$$\pi = \{(x^1, I^1), \ldots, (x^k, I^k)\}$$

where the I^j are left-closed intervals such that $\{I^1, \ldots, I^k\}$ is a partition of I and where $x^j \in \text{Cl}(I^j)$ $(1 \leq j \leq k)$.

Definition 2: If δ is a gauge of $\text{Cl}(I)$ (i.e. $\delta : \text{Cl}(I) \to (0, \infty)$), a P-partition $\pi = \{(x^1, I^1), \ldots, (x^k, I^k)\}$ is called δ-fine if

$$d(I^j) < \delta(x^j) \text{ for } j = 1, \ldots, k.$$

Now we define the concept of irregularity of a P-partition.

Definition 3: The rate of stretching $\sigma_0(I)$ of the left closed interval I is defined by

$$\sigma_0(I) = \max_{1\le i\le 2}(b_i - a_i) / \min_{1\le i\le 2}(b_i - a_i).$$

Moreover, if $\pi = \{(x^1, I^1), \ldots, (x^k, I^k)\}$ is a P-partition of I, we define

$$\sigma_1(x^j, I^j) = \frac{\max\{\mathrm{dist}(x^j, H_i^j) | i = 1, \ldots, 4 \text{ and } x^j \notin H_i^j\}}{\min\{\mathrm{dist}(x^j, H_i^j) | i = 1, \ldots, 4 \text{ and } x^j \notin H_i^j\}}$$

where $H_1^j, \ldots, H_4^j$ are the lines delimiting I^j.

As in [MAW2] we define the irregularity $\Sigma_0(\pi)$ and as in [MKH], we define another irregularity $\Sigma_1(\pi)$ of the P-partition π.

$$\Sigma_0(\pi) = \max_{1\le j\le k} \frac{\sigma_0(I^j)}{\sigma_0(I)}$$

and

$$\Sigma_1(\pi) = \max_{1\le j\le k} \frac{\sigma_1(x^j, I^j)}{\sigma_0(I)}.$$

Cousin's lemma: Let $\eta_0 = 0.4 \times 10^3$. If δ is a gauge of $\mathrm{Cl}(I)$, if $\eta \ge \eta_0$, we can find a δ-fine P-partition $\pi = \{(x^1, I^1), \ldots, (x^k, I^k)\}$ with x^j a vertex of I^j and $\Sigma_0(\pi) \le \eta$. It is easy to verify that in this case $\Sigma_0(\pi) = \Sigma_1(\pi)$.

Proof: See [BU].

Definition 4: If $\pi = \{(x^1, I^1), \ldots, (x^k, I^k)\}$ is a P-partition of I, if f is a real function on $\mathrm{Cl}(I)$, the sum associated with f is the real number $S(I, f, \pi)$, defined by

$$S(I, f, \pi) = \sum_{j=1}^{k} f(x^j) m(I^j),$$

where

$$m(I^j) = (b_2^j - a_2^j) \times (b_1^j - a_1^j) \text{ if } I^j = [a_1^j, b_1^j[\times [a_2^j, b_2^j[.$$

Definition 5: Let f be as in definition 4; we say that f is (GM)-integrable on I if there exists $J \in \mathbb{R}$ such that, for every $\varepsilon > 0$ and for each $\eta \ge \eta_0$ there exists a gauge δ on $\mathrm{Cl}(I)$ with the property that for every P-partition π δ-fine and η-regular ($\Sigma_1(\pi) \le \eta$), one has

$$|S(I, f, \pi) - J| \le \varepsilon.$$

For this integral, we have as usual a Cauchy criterion, a property of additivity and, in particular, a divergence theorem. (See [MKH].)

2. INTEGRALS OVER UNBOUNDED INTERVALS

In this section, I will denote an unbounded interval of the type:

$$I = I_1 \times I_2 \text{ where } I_j = [a_j, +\infty[\text{ or }]-\infty, b_j[\text{ or }]-\infty, +\infty[;$$

moreover for each real $r > 0$ and each vector $\vec{r} = (r_1, r_2)$ with $r_i > 0$ we consider

$$I_r = I \cap ([-r, r[\times[-r, r[)$$

and

$$I_{\vec{r}} = I \cap ([-r_1, r_1[\times[-r_2, r_2[).$$

Definition 6: Let f be a real function on $\mathrm{Cl}(I)$. We say that:

(1) f is (SGM)-integrable over I,

(2) f is (RGM)-integrable over I,

(3) f is (RRGM)-integrable over I, if there exists $J \in \mathbb{R}$ such that, for every $\varepsilon > 0$, for each $\eta \geq \eta_0$, there is a gauge δ on $\mathrm{Cl}(I)$ and a real $r_0 > 0$ satisfying:

(1) for every $r \geq r_0$ and for every P-partition π_r δ-fine, η-regular of I_r, one has

$$|S(I_r, f, \pi_r) - J| \leq \varepsilon,$$

(2) for every vector $\vec{r}$ with $r_i \geq r_0$, for every P-partition $\pi_{\vec{r}}$ δ-fine and η-regular of $I_{\vec{r}}$, one has

$$|S(I_{\vec{r}}, f, \pi_{\vec{r}}) - J)| \leq \varepsilon.$$

(3) for every vector $\vec{r}$ with $r_i \geq r_0$ and $\sigma_0(I_{\vec{r}}) \leq \eta$, the condition of (2) is satisfied.

In every case, if J exists, it is unique and we denote it (SGM)$\int_I f$, (RGM)$\int_I f$ or (RRGM)$\int_I f$.

Now, we will demonstrate the various properties of these integrals.

Proposition 1. If $\mathcal{S}(I)$, $\mathcal{R}(I)$ and $\mathcal{RR}(I)$ are the sets of the (SGM), (RGM) and (RRGM)-integrable functions on I, then

$$\mathcal{R}(I) \subset \mathcal{RR}(I) \subset \mathcal{S}(I).$$

Proof: The inclusion between $\mathcal{R}(I)$ and $\mathcal{RR}(I)$ is trivial. To prove the other, we must define three sets

$$\begin{aligned} \mathcal{A}_I &= \{i | i \in \{1, 2\} \text{ and } I_i = [a_i, +\infty[\} \\ \mathcal{B}_I &= \{i | i \in \{1, 2\} \text{ and } I_i =]-\infty, b_i[\} \\ \mathcal{C}_I &= \{i | i \in \{1, 2\} \text{ and } I_i =]-\infty, +\infty[\} \end{aligned}$$

and the number

$$\rho_0 = \max(\max_{i \in \mathcal{A}_I} |a_i|, \max_{i \in \mathcal{B}_I} |b_i|).$$

First, we will show that there is $\tilde{\rho}_0 \geq \rho_0$ such that for every $\rho \geq \tilde{\rho}_0$, $\sigma_0(I_\rho) \leq \eta_0$. We consider two cases. The first is when $\mathcal{C}_I$ is non-empty. If $\mathcal{C}_I = \{1, 2\}$, the result is trivial; otherwise, if $\rho > \rho_0$, then

$$\sigma_0(I_\rho) = \frac{2\rho}{\min(\min_{i \in \mathcal{A}_I}(\rho - a_i), \min_{i \in \mathcal{B}_I}(\rho + b_i))}.$$

Since $\sigma_0(I_\rho)$ tends to 2 when ρ tends to $+\infty$, it is easy to find $\tilde{\rho}_0 > \rho_0$ such that if $\rho > \tilde{\rho}_0$, $\sigma_0(I_\rho) \leq \eta_0$. The second case is when $\mathcal{C}_I$ is empty. Then, if $\rho > \rho_0$

$$\sigma_0(I_\rho) = \frac{\max(\max_{i\in A_I}(\rho - a_i),\ \max_{i\in B_I}(\rho + b_i))}{\min(\min_{i\in A_I}(\rho - a_i),\ \min_{i\in B_I}(\rho + b_i))}$$

Thus $\sigma_0(I_\rho)$ tends to 1 if ρ tends to $+\infty$ and we can conclude in the same way as above. Using this conclusion and the definition, it is easy to prove that if f is (RRGM)-integrable, f is (SGM)-integrable over I.

Proposition 2. The functional

$$\mathcal{F}:\ \mathcal{R}(I) \to I\!R;\ f \mapsto (\mathrm{RGM}) \int_I f$$

is linear and positive.

The same conclusion is also true for the (RRGM)-integral and the (SGM)-integral; the proof, being almost the same as that for the Perron integral [MAW1], is omitted.

The following property is the Cauchy criterion. Because it is similar for the three integrals, we quote and prove it only for the (SGM)-integral.

Proposition 3: Let f be a real function on $\mathrm{Cl}(I)$. Then f is (SGM)-integrable over I iff

(4) for every $\varepsilon > 0$, for each $\eta \geq \eta_0$ there is a gauge δ on $\mathrm{Cl}(I)$ and a real $r_0 > 0$ with the property that for every r, $r' \geq r_0$ and for every P-partition π_r, π'_r δ – fine and η – regular of I_r and $I_{r'}$,

$$|S(I_r,\ f,\ \pi_r) - S(I_{r'},\ f,\ \pi_{r'}| \leq \varepsilon.$$

Proof: The sufficient condition is trivial. For the necessary condition, let $\varepsilon > 0$ and $\eta \geq \eta_0$ be fixed. Taking $\varepsilon_k = 1/k$ in (4) we obtain a gauge δ_k and a real $r_{0,k}$ such that the inequality above holds. Without loss of generality, we can assume that

$$\delta_k \leq \delta_{k-1} \leq \ldots \leq \delta_1,$$

$$r_{0,k} \geq r_{0,k-1} \geq \ldots r_{0,1}.$$

Then, we can choose, for each $k \in I\!N_0$, a real $r_k \geq r_{0,k}$ and a δ_k-fine and η-regular P-partition π_k of I_{r_k}. If we denote by V_k the Riemann sum $S(I_{r_k}, f, \pi_k)$, $(V_k)_{k\in I\!N_0}$ forms a Cauchy sequence. Indeed, let $\varepsilon' > 0$ and $m_0 \in I\!N_0$ such that $\frac{1}{m_0} \leq \varepsilon'$. If k, $l \geq m_0$, then π_k and π_l are P-partitions δ_{m_0}-fine and η-regular of I_{r_k} and I_{r_l} (with r_k, $r_l \geq r_{0,m_0}$) and by (4)

$$|V_k - V_l| \leq \frac{1}{m_0} \leq \varepsilon'.$$

Therefore, $(V_k)_{k\in I\!N_0}$ converges, as k tends to infinity, to a limit J, say. Take $m_1 \in I\!N_0$ such that $\frac{2}{m_1} \leq \varepsilon$. We can choose $k_0 \geq m_1$ satisfying

$$|V_k - J| \leq \frac{1}{m_1} \text{ for every } k \geq k_0.$$

The gauge δ and the real $r_0 > 0$ that we need in the definition (3) of the (SGM)-integral are δ_{k_0} and r_{0,k_0}. Indeed, letting $r \geq r_{0,k_0}$ and $\tilde{\pi}_r$ be a P-partition δ_{k_0}-fine and η-regular of I_r

$$\begin{aligned} |S(I_r, f, \tilde{\pi}_r) - J| &\leq |S(I_r, f, \tilde{\pi}_r) - V_{k_0}| + |V_{k_0} - J| \\ &\leq 1/k_0 + 1/m_1 \leq 2/m_1 \leq \varepsilon. \end{aligned}$$

As a consequence of the Cauchy criterion, there are some corollaries.

Corollary 1: If f is (SGM)-integrable over I, then f is (GM)-integrable over I_c for every $c > 0$. Moreover,

$$\lim_{c \to +\infty} (\text{GM}) \int_{I_c} f \text{ exists and is equal to (SGM)} \int_I f.$$

Proof: First we want to show that f is (GM)-integrable on I_c. By the Cauchy criterion for the (GM)-integral [MKH], it is sufficient to prove that for every $\varepsilon > 0$, for each $\eta \geq \eta_0$ there is a gauge δ on $\text{Cl}(I_c)$ satisfying, for every P-partition π, π' δ-fine, η-regular of I_c,

$$(5) \quad |S(I_c, f, \pi) - S(I_c, f, \pi')| \leq \varepsilon.$$

Let $\varepsilon > 0$, $\eta \geq \eta_0$ be fixed and consider $\eta^* = \eta\sigma_0(I_c)$. There is a gauge δ^* on $\text{Cl}(I)$ and a real $r_0 > 0$ such that the condition (4) holds. Without loss of generality, we can suppose that

$$\begin{aligned} \delta^*(x) &\leq \tfrac{1}{2}\text{dist}(x, \text{Cl}(I_c)) \text{ if } x \notin \text{Cl}(I_c) \\ &\leq \tfrac{1}{2}Q(x, I_c) \text{ if } x \in \text{Cl}(I_c), \end{aligned}$$

where

$$Q(x, I_c) = \min\{\text{dist}(x, H_c^i) | i = 1, \ldots, 4 \text{ and } x \notin H_c^i\},$$

H_c^i being the lines which form the boundary of I_c. Therefore, it is sufficient to take $\delta = \delta^*|_{\text{Cl}(I_c)}$ in (5). Indeed, if $c \geq r_0$, then by (4), the condition (5) holds immediately. If $c < r_0$, let π^* be a P-partition δ^*-fine and η-regular of I_{r_0},

$$\pi^* = \{(y^1, J^1), \ldots, (y^p, J^p)\}.$$

Since $c < r_0$, $I_c \subseteq I_{r_0}$ and by the choice of the gauge δ, if $j \in \{1, \ldots, p\}$ is such that

$$\text{int}(J^j \cap (I_{r_0} \setminus I_c)) \neq \emptyset$$

then

$$J^j \cap I_{r_0} \setminus I_c = \cup_{k=1}^{m_j} K_k^j$$

where K_k^j is a left closed interval with, if $z_k^j = y^j$,

$$\sigma_1(z_k^j,\, K_k^j) \leq \sigma_1(y^j,\, J^j) \text{ and } d(K_k^j) < \delta^*(y^j).$$

We define

$$\mathcal{J} = \{j | 1 \leq j \leq p \text{ with int } (J^j \cap (I_{r_0} \setminus I_c)) \neq \emptyset\}$$

and

$$\pi^*|_{I_{r_0} \setminus I_c} = \{(z_k^j,\, K_k^j) |\ 1 \leq k \leq m_j,\ j \in \mathcal{J}\}.$$

Now let $\pi = \{(x_0^1,\, I_0^1), \ldots, (x_0^s,\, I_0^s)\}$ and $\pi' = \{(x_1^1,\, I_1^1), \ldots, (x_1^t,\, I_1^t)\}$ be two P-partitions δ-fine and η-regular of I_c. It is easy to verify that $\pi \cup (\pi^*|_{I_{r_0} \setminus I_c})$ and $\pi' \cup (\pi^*|_{I_{r_0} \setminus I_c})$ are two P-partitions δ-fine and η^*-regular of I_{r_0}. Therefore,

$$|S(I_{r_0},\, f,\, \pi \cup (\pi^*|_{I_{r_0} \setminus I_c})) - S(I_{r_0},\, f,\, \pi' \cup (\pi^*|_{I_{r_0} \setminus I_c}))| \leq \varepsilon,$$

i.e.

$$|S(I_c,\, f,\, \pi) - S(I_c,\, f,\, \pi')| \leq \varepsilon.$$

The second step is to prove that $\lim_{c \to +\infty} \mathrm{GM} \int_{I_c} f$ exists and is equal to the (SGM)-integral of f. Let $\varepsilon > 0$ be fixed and, taking $\eta = \eta_0$, there is a gauge δ on $\mathrm{Cl}(I)$ and a real $r_0 > 0$ such that (1) is satisfied, and f being (GM)-integrable on I_c, there is also a gauge $\tilde{\delta}_c$ on $\mathrm{Cl}(I_c)$ such that if $\tilde{\pi}$ is a P-partition $\tilde{\delta}_c$-fine and η-regular of I_c, one has

(6) $|S(I_c,\, f,\, \tilde{\pi}) - (\mathrm{GM}) \int_{I_c} f| \leq \varepsilon.$

Let $c_0 = r_0$ and assume that $c \geq c_0$. To prove the corollary, we must consider the gauge on $\mathrm{Cl}(I_c)$, $\delta^* = \min(\delta|_{\mathrm{Cl}(I_c)}, \tilde{\delta}_c)$ and π^* a P-partition δ^*-fine and η-regular of I_c. Indeed, by (1) and (6)

$$\begin{aligned} |(\mathrm{GM}) \textstyle\int_{I_c} f - (\mathrm{SGM}) \int_I f| &\leq |(\mathrm{GM}) \textstyle\int_{I_c} f - S(I_c,\, f,\, \pi^*)| \\ &\quad + |S(I_c,\, f,\, \pi^*) - (\mathrm{SGM}) \textstyle\int_I f| \\ &\leq 2\varepsilon. \end{aligned}$$

In the same way, we can prove the following corollaries.

Corollary 2: If f is (RRGM)-integrable over I, then f is (GM)-integrable over $I_{\vec{c}}$ for every vector $\vec{c}$ with $c_i > 0$, and for each $\eta \geq \eta_0$

$$\lim_{\substack{\vec{c} \rightsquigarrow +\infty \\ \sigma_0(I_{\vec{c}}) \leq \eta}} (\mathrm{GM}) \int_{I_{\vec{c}}} f \text{ exists and equals (RRGM)} \int_I f.$$

Remark: We write

$$\lim_{\substack{\vec{c} \rightsquigarrow +\infty \\ \sigma_0(I_{\vec{c}}) \leq \eta}} (\mathrm{GM}) \int_{I_{\vec{c}}} f = J$$

if for every $\varepsilon > 0$ there is a real $c_0 > 0$ such that for each vector $\vec{c} = (c_1, c_2)$ with $c_i \geq c_0$ and $\sigma_0(I_{\vec{c}}) \leq \eta$,

$$|(\mathrm{GM}) \int_{I_{\vec{c}}} f - J| \leq \varepsilon.$$

Corollary 3: If f is (RGM)-integrable over I, then f is (GM)-integrable over $I_{\vec{c}}$ for each vector $\vec{c}$ with $c_i > 0$ and

$$\lim_{\vec{c} \leadsto +\infty} (\mathrm{GM}) \int_{I_{\vec{c}}} f = (\mathrm{RGM}) \int_I f.$$

We would like now to prove converses of corollaries 1 and 2 because this would generalise Hake's theorem to two dimensions. (For Hake's theorem, see [MAW1].)

Theorem 1. If f is a real function on $\mathrm{Cl}(I)$ which is (GM)- integrable on $I_{\vec{c}}$ for every $\vec{c}$ with $c_i > 0$, if for each $\eta \geq \eta_0$,

$$\lim_{\substack{\vec{c} \leadsto +\infty \\ \sigma_0(I_{\vec{c}}) \leq \eta}} (\mathrm{GM}) \int_{I_{\vec{c}}} f \text{ exists}$$

and is equal to J, then f is (RRGM)-integrable over I and (RRGM) $\int_I f = J$.

Proof: Let $\eta \geq \eta_0$ and $\varepsilon > 0$ be fixed. By assumptions, there is $c_0 > 0$ such that for every $\vec{c}$ with $c_i \geq c_0$ and $\sigma_0(I_{\vec{c}}) \leq \eta$, we have

(7) $|(\mathrm{GM}) \int_{I_{\vec{c}}} f - J| \leq \varepsilon.$

Let $\mathcal{A}_I$, $\mathcal{B}_I$, $\mathcal{C}_I$ and ρ_0 be as in Proposition 1. We can assume, without loss of generality, that

$$\mathcal{A}_I = \{1\} \text{ and } \mathcal{C}_I = \{2\}.$$

Take $r_0 = \max(\rho_0, c_0)$ and $\vec{r}$ a vector satisfying $r_i \geq r_0$ and $\sigma_0(I_{\vec{r}}) \leq \eta$, then $I_{\vec{r}} = [a_1, r_1[\times[-r_2, r_2[$. Furthermore we define

$$a_1^{j_1} = a_1 + j_1, \ I_1^{j_1} = [a_1^{j_1}, a_1^{j_1+1}[,$$

$$c_2^{j_2} = j_2, \ I_2^{j_2} = [c_2^{j_2}, c_2^{j_2+1}[,$$

$$J_2^{j_2} = [-c_2^{j_2+1}, -c_2^{j_2}[,$$

$$I_0^{j_1 j_2} = I_1^{j_1} \times I_2^{j_2}, \ I_1^{j_1 j_2} = I_1^{j_1} \times J_2^{j_2} \ (j_1, j_2 \geq 0).$$

Since f is (GM)-integrable over $I_k^{j_1 j_2}$, there is a gauge $\delta_k^{j_1 j_2}$ on $\mathrm{Cl}(I_k^{j_1 j_2})$ such that

$$|S(I_k^{j_1 j_2}, f, \pi_k^{j_1 j_2}) - (\mathrm{GM}) \int_{I_k^{j_1 j_2}} f| \leq \frac{\varepsilon}{2^{j_1+j_2+2}}$$

for every P-partition $\pi_k^{j_1 j_2}$ $\delta_k^{j_1 j_2}$-fine and η^2-regular of $I_k^{j_1 j_2}$, where $k = 0$ or 1.

These gauges allow us to define a gauge δ on $\mathrm{Cl}(I)$. Indeed, if

$$\begin{aligned} x &\in \mathrm{int}(I_k^{j_1 j_2}),\ \delta(x) = \min(\delta_k^{j_1 j_2}(x),\ Q(x,\ I_k^{j_1 j_2})), \\ x &\in \mathrm{Cl}(I_k^{j_1 j_2}) \cap \mathrm{Cl}(I_{k'}^{j_1' j_2'}), \end{aligned}$$

then

$$\delta(x) = \min(\delta_k^{j_1 j_2}(x),\ \delta_{k'}^{j_1' j_2'}(x), Q(x,\ I_k^{j_1 j_2}),\ Q(x,\ I_k^{\prime j_1' j_2'}))$$

where $(j_1,\ j_2) \neq (j_1',\ j_2')$ and $k,\ k' = 0$ or 1.

To prove the theorem, we must show that if $\vec{r}$ is a vector with $r_i \geq r_0$ and $\sigma_0(I_{\vec{r}}) \leq \eta$, if $\pi_{\vec{r}} = \{(y^1,\ K^1), \ldots, (y^p,\ K^p)\}$ is a P-partition δ-fine and η-regular of $I_{\vec{r}}$ one has

$$|S(I_{\vec{r}},\ f,\ \pi_{\vec{r}}) - J| \leq \varepsilon.$$

But by the choice of the gauge, if

$$\begin{aligned} y^t &\in \mathrm{int}(I_k^{j_1 j_2}),\ K^t \subset I_k^{j_1 j_2} \text{ and} \\ y^t &\in \mathrm{Cl}(I_k^{j_1 j_2}) \cap \mathrm{Cl}(I_{k'}^{j_1' j_2'}) \text{ with } (j_1',\ j_2') \neq (j_1,\ j_2) \text{ and } k,\ k' = 0 \text{ or } 1, \end{aligned}$$

then

$$\begin{aligned} K^t &\subset \mathrm{Cl}(I_k^{j_1 j_2}) \cup \mathrm{Cl}(I_{k'}^{j_1' j_2'}) \text{ and} \\ m(K^t) &= m(K^t \cap I_k^{j_1 j_2}) + m(K^t \cap I_{k'}^{j_1' j_2'}). \end{aligned}$$

To continue, we must define two integers $q_1,\ q_2$ and some intervals associated with these integers: q_1 is the greatest positive integer such that $a_1^{q_1} \leq r_1$ and $\tilde{I}_1^{q_1} = [a_1^{q_1},\ r_1[$, q_2 is the greatest positive integer such that $c_2^{q_2} \leq r_2$ and $\tilde{I}_2^{q_2} = [c_2^{q_2},\ r_2[$, $\tilde{J}_2^{q_2} = [-r_2,\ -c_2^{q_2}[$. On the one hand, we consider for $0 \leq j_i \leq q_i - 1$ and $k = 0$ or 1, the P-partition $\pi|_{I_k^{j_1 j_2}}$ defined by

$$\pi|_{I_k^{j_1 j_2}} = \{(z_k^{l j_1 j_2},\ K_k^{l j_1 j_2}) | l \in \mathcal{H}_k^{j_1 j_2}\}$$

where $\mathcal{H}_k^{j_1 j_2} = \{j | j \in \{1, \ldots, p\}$ and $\mathrm{int}(K^j \cap I_k^{j_1 j_2}) \neq \emptyset\}$ and if $l \in \mathcal{H}_k^{j_1 j_2}$, $K_k^{l j_1 j_2} = K^l \cap I_k^{j_1 j_2}$ and $z_k^{l j_1 j_2} = y^l$. It is easy to prove, by the remark above, that $\pi|_{I_k^{j_1 j_2}}$ is a P-partition $\delta_k^{j_1 j_2}$-fine and η^2-regular of $I_k^{j_1 j_2}$. Thus,

$$|S(I_k^{j_1 j_2},\ f,\ \pi|_{I_k^{j_1 j_2}} - (\mathrm{GM}) \int_{I_k^{j_1 j_2}} f| \leq \frac{\varepsilon}{2^{j_1 + j_2 + 2}},$$

for $0 \leq j_1 \leq q_1 - 1$, $0 \leq j_2 \leq q_2 - 1$ and $k = 0$ or 1. On the other hand, we consider the following intervals

$$\begin{aligned} \tilde{I}_0^{j_1 q_2} &= I_1^{j_1} \times \tilde{I}_2^{q_2},\ \tilde{I}_1^{j_1 q_2} = I_1^{j_1} \times \tilde{J}_2^{q_2}, \\ \tilde{I}_0^{q_1 j_2} &= \tilde{I}_1^{q_1} \times I_2^{j_2},\ \tilde{I}_1^{q_1 j_2} = \tilde{I}_1^{q_1} \times J_2^{j_2}, \\ \tilde{I}_0^{q_1 q_2} &= \tilde{I}_1^{q_1} \times \tilde{I}_2^{q_2},\ \tilde{I}_1^{q_1 q_2} = \tilde{I}_1^{q_1} \times \tilde{J}_2^{q_2}. \end{aligned}$$

We can verify that π restricted, as above, to each of these intervals is a P-partition $\delta_0^{j_1 q_2}$ ($\delta_1^{j_1 q_2}, \ldots, \delta_1^{q_1 q_2}$ respectively)-fine and η^2-regular of $\tilde{I}_0^{j_1 q_2}$ ($\tilde{I}_1^{j_1 q_2}, \ldots, \tilde{I}_1^{q_1 q_2}$

respectively) and by the lemma of Saks-Henstock for the (GM)-integral [MKH], one has

$$\begin{aligned}
&|(\mathrm{GM})\textstyle\int_{\tilde{I}_0^{j_1 q_2}} f - S(\tilde{I}_0^{j_1 q_2},\, f,\, \pi|_{\tilde{I}_k^{j_1 q_2}})| \;\leq\; \frac{\varepsilon}{2^{j_1+q_2+2}},\\
&\vdots\\
&|(\mathrm{GM})\textstyle\int_{\tilde{I}_1^{q_1 q_2}} f - S(\tilde{I}_1^{q_1 q_2},\, f,\, \pi|_{\tilde{I}_k^{q_1 q_2}})| \;\leq\; \frac{\varepsilon}{2^{q_1+q_2+2}}.
\end{aligned}$$

Finally, by the additivity of the (GM)-integral,

$$\begin{aligned}
|S(I_{\vec{r}},\, f,\, \pi) - J| &\leq |S(I_{\vec{r}},\, f,\, \pi) - (\mathrm{GM})\textstyle\int_{I_{\vec{r}}} f| + |(\mathrm{GM})\textstyle\int_{I_{\vec{r}}} f - J|\\
&\leq \textstyle\sum_{k=0}^{1}[\sum_{j_1=0}^{q_1-1}\sum_{j_2=0}^{q_2-1} |S(I_k^{j_1 j_2},\, f,\, \pi|_{I_k^{j_1 j_2}}) - (\mathrm{GM})\int_{I_k^{j_1 j_2}} f|\\
&\quad + \textstyle\sum_{j_1=0}^{q_1-1} |S(\tilde{I}_k^{j_1 q_2},\, f,\, \pi|_{\tilde{I}_k^{j_1 q_2}}) - (\mathrm{GM})\int_{\tilde{I}_k^{j_1 q_2}} f|\\
&\quad + \textstyle\sum_{j_2=0}^{q_2-1} |S(\tilde{I}_k^{q_1 j_2},\, f,\, \pi|_{\tilde{I}_k^{q_1 j_2}}) - (\mathrm{GM})\int_{\tilde{I}_k^{q_1 j_2}} f|\\
&\quad + |S(\tilde{I}_k^{q_1 q_2},\, f,\, \pi|_{\tilde{I}_k^{q_1 q_2}}) - (\mathrm{GM})\textstyle\int_{\tilde{I}_k^{q_1 q_2}} f|] + (\mathrm{GM})\textstyle\int_{I_{\vec{r}}} f - J|\\
&\leq 2\varepsilon \textstyle\sum_{j_1=0}^{q_1}\sum_{j_2=0}^{q_2} \frac{1}{2^{j_1+j_2+2}} + |(\mathrm{GM})\int_{I_{\vec{r}}} f - J|\\
&= 2\varepsilon \textstyle\sum_{j_1=0}^{q_1} \frac{1}{2^{j_1+1}} \sum_{j_2=0}^{q_2-1} \frac{1}{2^{j_2+1}} + |(\mathrm{GM})\int_{I_{\vec{r}}} f - J|\\
&\leq 3\varepsilon \text{ by}(7).
\end{aligned}$$

For the (SGM)-integral, there is, as we have said above, a similar theorem. We state this theorem but omit the proof.

Theorem 2: Let $f : f \to I\!R$ be a function (GM)-integrable over I_c for every $c > 0$. If $\lim_{c\to+\infty}(\mathrm{GM})\int_{I_c} f$ exists and is equal to J then f if (GM)-integrable over I and

$$(\mathrm{SGM})\int_I f = J.$$

At present, we have no proof for the (RGM)-integral because when $\vec{r}$ is any vector $\sigma_0(I_{\vec{r}})$ is uncontrollable. On the other hand due to Theorems 1 and 2, we can obtain some properties of additivity for these integrals.

Proposition 4. Let $I = \cup_{j=1}^{t} K_j$ where K_j are intervals of the same type as I and moreover with $\mathrm{int}(K_j) \cap \mathrm{int}(K_l) = \emptyset$ if $j \neq l$. If f is a real function on $\mathrm{Cl}(I)$ (SGM)-integrable over each K_j, then f is (SGM)-integrable over I and

$$(\mathrm{SGM})\int_I f = \sum_{j=1}^{t} (\mathrm{SGM})\int_{K_j} f$$

Proof: Since f is (SGM)-integrable over each K_j, f is (GM)-integrable over $(K_j)_c$ for every $c > 0$ and $\lim_{c\to+\infty}(\mathrm{GM})\int_{(K_j)_c} f = (\mathrm{SGM})\int_{K_j} f$. But $I = \cup_{j=1}^{t} K_j$, thus

$$\begin{aligned}
I_c &= (\cup_{j=1}^{t} K_j) \cap [-c,\, c[\times[-c,\, c[,\\
&= \cup_{j=1}^{t} (K_j)_c \text{ with } \mathrm{int}((K_j)_c) \cap\ \mathrm{int}((K_l)_c) = \emptyset, \text{ if } j \neq l.
\end{aligned}$$

Therefore, by the additivity of the (GM)-integral

$$(\mathrm{GM})\int_{I_c} f = \sum_{j=1}^{t} (\mathrm{GM})\int_{(K_j)_c} f$$

And, if we take the limit on c,

$$\lim_{c\to+\infty} (\mathrm{GM}) \int_{I_c} f = \sum_{j=1}^{t} \lim_{c\to+\infty} (\mathrm{GM}) \int_{(K_j)_c} f = \sum_{j=1}^{t} (\mathrm{SGM}) \int_{K_j} f.$$

Consequently, by theorem 2, f is (SGM)-integrable and

$$(\mathrm{SGM}) \int_I f = \sum_{j=1}^{t} (\mathrm{SGM}) \int_{K_j} f.$$

In the same way, we have the following proposition.

Proposition 5. Let $I = \cup_{j=1}^{t} K_j$ where K_j is as in Proposition 4. If f is a real function on $\mathrm{Cl}(I)$, (RGM)-integrable over each K_j, then f is (RRGM)-integrable on I and moreover

$$(\mathrm{RRGM}) \int_I f = \sum_{j=1}^{t} (\mathrm{RGM}) \int_{K_j} f.$$

3. CONVERGENCE OF DOUBLE SERIES

In this section, I represents the interval $[1, +\infty[\times[1, +\infty[$ and $\sum_{i,j} a_{ij}$ a real double series.

Definition 7. The partial sum of $\sum_{i,j} a_{ij}$ is

$$S_{mn} = \sum_{i=1}^{m} \sum_{j=1}^{n} a_{ij}$$

and we define also

$$\tilde{\sigma}_0(m, n) = \frac{\max(m, n)}{\min(m, n)}.$$

$$\tilde{\sigma}_0(m, n) = \sigma_0([1, m+1[\times[1, n+1[).$$

We consider for the double series three types of convergence ([ASH]).

Definition 8. We say that

(8) $\sum_{i,j} a_{ij}$ S-converges to A (Square-convergence)

(9) $\sum_{i,j} a_{ij}$ R-converges to A (Rectangular-convergence)

if for every $\varepsilon > 0$, there is $n_0 \in \mathbb{N}_0$ satisfying

(8) for each $n \geq n_0$, $|S_{nn} - A| \leq \varepsilon$,

(9) for each $m,\ n \geq n_0$, $|S_{mn} - A| \leq \varepsilon$, and

(10) $\sum_{i,j} a_{ij}$ RR-converges to A (Restricted Rectangular convergence)

if, for every $\varepsilon > 0$, for each $\eta > 1$, there exists $n_0 \in \mathbb{N}_0$ such that

(10) if $m,\ n \geq n_0$ with $\tilde{\sigma}_0(m,\ n) \leq \eta$, $|S_{mn} - A| \leq \varepsilon$.

Remark 1 : The R-convergence of a series implies its RR-convergence which implies the S-convergence of this series.

Remark 2 : We give here a representation of couples (m, n) which satisfy $m,\ n \geq n_0$ and $\tilde{\sigma}_0(m,\ n) < \eta$ with for example $n_0 = 1$ and $\eta = 2$.

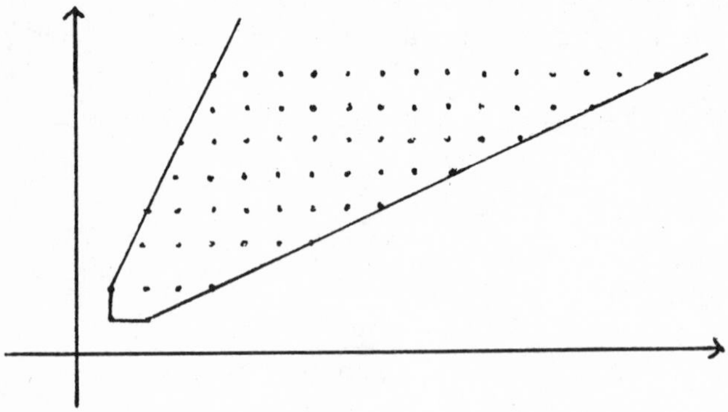

Remark 3 : The condition $\tilde{\sigma}_0(m,\ n) \leq \eta$ is equivalent to

$$\frac{n}{\eta} \leq m \leq \eta n.$$

We now state two lemmas that we need to establish a connection between integrability and convergence of double series but first we associate with each double series $\sum_{i,j} a_{ij}$ a function on I defined by

$$f(x,\ y) = a_{ij} \text{ if } x \in [i,\ i+1[\text{ and } y \in [j,\ j+1[.$$

Lemma 1 : (Without proof.) The function f, defined above, is (GM)-integrable over $[1,\ c_1[\times[1,\ c_2[$ for every $c_1,\ c_2 > 1$ and moreover

$$\begin{aligned}(GM)\int_{[1,c_1[\times[1,c_2[} f \;=\; & \sum_{i=1}^{c_1^*-1}\sum_{j=1}^{c_2^*-1} a_{ij} + \sum_{j=1}^{c_2^*-1} a_{c_1^* j}(c_1 - c_1^*) \\ & + \sum_{i=1}^{c_1^*-1} a_{i c_2^*}(c_2 - c_2^*) + a_{c_1^* c_2^*}(c_1 - c_1^*)(c_2 - c_2^*)\end{aligned}$$

where c_i^* is the integer part of c_i.

Lemma 2. If c_1, $c_2 \geq 3$ are such that $\sigma_0([1, c_1[\times[1, c_2[) \leq \eta$ with $\eta \geq \eta_0$

$$\begin{aligned} \tilde{\sigma}_0(c_1^* - 1,\ c_2^* - 1) &\leq 2\eta, \\ \tilde{\sigma}_0(c_1^* - 1,\ c_2^*) &\leq 2\eta, \\ \tilde{\sigma}_0(c_2^* - 1,\ c_1^*) &\leq 2\eta, \\ \tilde{\sigma}_0(c_1^*,\ c_2^*) &\leq 2\eta \end{aligned}$$

Proof. By symmetry, we can assume that $c_1 \geq c_2$. Thus,

(11) $\sigma_0([1, c_1[\times[1, c_2[) \leq \eta$ iff $(c_1 - 1) \leq \eta(c_2 - 1)$.

But

$$c_1 - 1 = c_1^* - 1 + \gamma \text{ with } 0 \leq \gamma < 1$$

and

$$c_2 - 1 = c_2^* - 1 + \mu \text{ with } 0 \leq \mu < 1.$$

By (11), this implies that

$$(c_1^* - 1) \leq \eta(c_2^* - 1 + \mu).$$

Therefore, one has

(12) $(c_1^* - 1) \leq \eta(c_2^* - 1 + 1) \leq 2\eta(c_2^* - 1)$.

and

$$c_1^* \leq \eta(c_2^* - 1 + 2) \leq 2\eta(c_2^* - 1).$$

But then, $c_1^* \leq 2\eta c_2^*$. It remains to show that $\tilde{\sigma}_0(c_1^* - 1,\ c_2^*) \leq 2\eta$. If $c_1^* = c_2^*$, $\tilde{\sigma}_0(c_1^* - 1,\ c_2^*) = \frac{c_1^*}{c_1^* - 1}$ and then, for $\tilde{\sigma}_0(c_1^* - 1,\ c_2^*) \leq 2\eta$, it is sufficient that $c_1^* \geq \frac{2\eta_0}{2\eta_0 - 1}$. This is the case. If $c_1^* > c_2^*$, $\tilde{\sigma}_0(c_1^* - 1,\ c_2^*) = \frac{c_1^* - 1}{c_2^*}$ and the result follows from (12).

Theorem 3 $\sum_{i,j} a_{ij}$ is RR-convergent iff f is (RRGM)-integrable on I.

Proof. Assume first that $\sum_{i,j} a_{ij}$ RR-converges to A. We must show that for each $\eta \geq \eta_0$

$$\lim_{\substack{\tilde{\varepsilon} \rightsquigarrow +\infty \\ \sigma_0(I_{\tilde{\varepsilon}}) \leq \eta}} (\text{GM}) \int_{I_{\tilde{\varepsilon}}} f$$

exists and is equal to A because then, by Theorem 1, we can conclude that f if (RRGM)-integrable. Let $\eta \geq \eta_0$ and $\varepsilon > 0$ be fixed. By the RR-convergence, we find $n_0 \in \mathbb{N}_0$ such that (10) holds with $\eta' = 2\eta$ and $\varepsilon' = \varepsilon/9$. Furthermore, choose $c_0 = \max(n_0 + 1, 3)$. If c_1, $c_2 \geq c_0$ and $\sigma_0(I_{\tilde{\varepsilon}}) \leq \eta$, by Lemma 1, f is (GM)-integrable over $[1, c_1[\times[1, c_2[$ and moreover

$$
\begin{aligned}
|(GM) \int_{[1,c_1[\times[1,c_2[} f - A| &\leq |S_{c_1^*-1, c_2^*-1} - A| + \sum_{j=1}^{c_2^*-1} |a_{c_1^* j}(c_1 - c_1^*)| \\
&\quad + \sum_{j=1}^{c_2^*-1} |a_{j c_2^*}(c_2 - c_2^*)| + |a_{c_1^* c_2^*}(c_1 - c_1^*)(c_2 - c_2^*)| \\
&\leq |S_{c_1^*-1, c_2^*-1} - A| + |S_{c_1^*, c_2^*-1} - S_{c_1^*-1, c_2^*-1}| \\
&\quad + |S_{c_1^*-1, c_2^*} - S_{c_1^*-1, c_2^*-1}| \\
&\quad + |S_{c_1^*, c_2^*} - S_{c_1^*, c_2^*-1} - S_{c_1^*-1, c_2^*} + S_{c_1^*-1, c_2^*-1}| \\
&\leq \varepsilon
\end{aligned}
$$

by (10) and Lemma 2. So,

$$
\lim_{\substack{\tilde{\varepsilon} \to +\infty \\ \sigma_0(I_{\tilde{\varepsilon}}) \leq \eta}} (GM) \int_{I_{\tilde{\varepsilon}}} f = A.
$$

Now, suppose that f is (RRGM)-integrable. We want to show that $\sum_{i,j} a_{ij}$ RR-converges to (RRGM)$\int_I f$. Let $\varepsilon > 0$ and $\eta > 1$ be fixed. On the one hand, assume that $1 < \eta \leq \eta_0$, then there is $c_0 > 0$ such that if c_1, $c_2 \geq c_0$ and $\sigma_0([1, c_1[\times[1, c_2[) \leq \eta_0$,

$$
|(GM) \int_{[1,c_1[\times[1,c_2[} f - (RRGM) \int_I f| \leq \varepsilon.
$$

Taking $n_0 = c_0^*$ it follows that if n, $m \geq n_0$ and $\tilde{\sigma}_0(m, n) \leq \eta$

$$
|(GM) \int_{[1,n+1[\times[1,m+1[} f - (RRGM) \int_I f| \leq \varepsilon.
$$

Then, by lemma 1,

$$
|S_{mn} - (RRGM) \int_I f| \leq \varepsilon.
$$

On the other hand, when $\eta \geq \eta_0$ there exists again a real $c_0 > 0$ such that if c_1, $c_2 \geq c_0$ and $\sigma_0([1, c_1[\times[1, c_2[) \leq \eta$

$$
|(GM) \int_{[1,c_1[\times[1,c_2[} f - (RRGM) \int_I f| \leq \varepsilon.
$$

and this completes the proof.

Proposition 4 : If f is (SGM)-integrable (resp. (RGM)-integrable) over I, then $\sum_{i,j} a_{ij}$ is S-convergent (R-convergent) to (SGM) $\int_I f$ ((RGM) $\int_I f$).

Proof: This can be proved in a way similar to the proof of the second part of theorem 3.

The conclusion of Theorem 3, Proposition 4 and Remark 1 is the following diagram.

$\sum_{i,j} a_{ij}$ R-converges $\longrightarrow$ $\sum_{i,j} a_{ij}$ RR-converges $\longrightarrow$ $\sum_{i,j} a_{ij}$ S-converges

f is (RGM)-integrable $\longrightarrow$ f is (RRGM)-integrable $\longrightarrow$ f is (SGM)-integrable

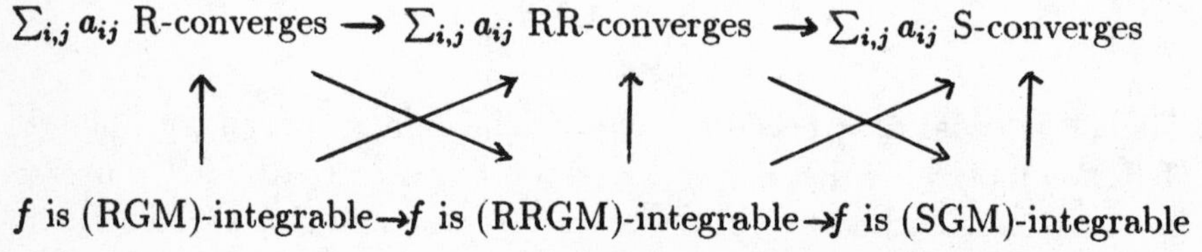

The implications which do not appear in the diagram are false, except one: the implication between the R-convergence and the (RGM)-integrability of f. At present, we know neither how to prove this implication nor how to find a counterexample. This remains an open problem. On the other hand, we give some counterexamples of the implications which are false.

Counterexample 1. Take a_{ij} as follows:

$$\begin{aligned} a_{jj} &= 1-j \\ a_{j1} &= j-1 \\ a_{ij} &= 0 \text{ otherwise.} \end{aligned}$$

Then $S_{nn} = 0$ and $\sum_{i,j} a_{ij}$ S-converges to 0. But f is not (SGM)-integrable over I. Indeed, if f were (SGM)-integrable over I, we should have that

$$(\mathrm{SGM})\int_I f = 0,$$

But then, by corollary 1, we would have

$$\lim_{c\to+\infty} (\mathrm{GM})\int_{[1,c[\times[1,c[} f = 0$$

This is impossible because if we take $\varepsilon = \frac{1}{8}$ for every $c_0 > 0$, we can find $c \geq c_0$ such that

$$|(\mathrm{GM})\int_{[1,c[\times[1,c[} f| \geq \varepsilon.$$

It is enough to take $c = c_0^* + \frac{5}{2}$

Counterexample 2. There exists some series RR-convergent but for which f is not (RGM)-integrable over I; for instance, the following series

$$a_{j^2+1,1} = j;\ a_{j^2+1,j+1} = -j \text{ and } a_{ij} = 0 \text{ otherwise.}$$

This is a RR-convergent series (to 0), but f cannot be (RGM)-integrable because by Proposition 4, the series should be R-convergent and this is impossible because $S_{j^2+1,j} = j$

Counterexample 3. It is possible to find a series $\sum_{i,j} a_{ij}$ which is not RR-convergent but for which f is (SGM)-integrable over I. Indeed, consider the series $\sum_{i,j} a_{ij}$ where

$$\begin{aligned} a_{ij} &= -j \text{ if } i-j=1, \\ &= +j \text{ if } j-i=1, \\ &= 0 \text{ otherwise.} \end{aligned}$$

By Theorem 2, since $(\mathrm{GM})\int_{[1,c[\times[1,c[} f = 0$, f is SGM-integrable and

$$(\mathrm{SGM})\int_I f = 0$$

But $\sum_{i,j} a_{ij}$ is not RR-convergent because if it were RR-convergent then f would be (RRGM)-integrable with its (RRGM)-integral equal to 0 and this is impossible because

$$\lim_{\substack{\varepsilon \rightsquigarrow +\infty \\ \sigma_0([1,c_1[\times[1,c_2[)\leq\eta_0}} (\mathrm{GM}) \int_{[1,c_1[\times[1,c_2[} f \neq 0.$$

Indeed, taking $\varepsilon = 1$, for every $c_0 > 1$, we can find $c_1, c_2 \geq c_0$ with $\sigma_0([1,c_1[\times[1,c_2[) \leq \eta_0$ and such that

$$|(\mathrm{GM}) \int_{[1,c_1[\times[1,c_2[} f| > 1.$$

It is sufficient to take $c_1 = c_0^* + 2$ and $c_2 = 2c_0^* + 2$ because we then have

$$|(\mathrm{GM}) \int_{[1,c_1[\times[1,c_2[} f| = 1 + c_0^* \geq 1.$$

Counterexample 4. Let $\sum_{i,j} a_{ij}$ be a double series with f (RRGM)-integrable, $\sum_{i,j} a_{ij}$ is not necessarily R-convergent. For instance, take the series of counterexample 2; f is (RRGM)-integrable over I because $\sum_{i,j} a_{ij}$ RR-converges but $\sum_{i,j} a_{ij}$ is not R-convergent.

Counterexample 5. The series of counterexample 3 is also a counterexample of the implication between the S-convergence and the (RRGM)-integrability of f.

The last result is the following.

Proposition 5. Let $\sum_{i,j} a_{ij}$ be a series with $a_{ii} = 0$, for every i or with $a_{ij} \geq 0$ for every i, j. $\sum_{i,j} a_{ij}$ S-converges iff f is SGM-integrable on I.

Proof : We need only prove the case of $a_{ii} = 0$ because when $a_{ij} \geq 0$, f is a positive function and for such functions the result is already known (and easily demonstrable) since the (SGM)-integral is equivalent to the Lebesgue integral. If f is (SGM)-integrable, then $\sum_{i,j} a_{ij}$ is certainly S-convergent by Proposition 4. Now, assume that $\sum_{i,j} a_{ij}$ S-converges to A. Therefore if $\varepsilon > 0$ is fixed, there is $n_0 \in I\!N_0$ with the property that the condition (8) holds. Take $c_0 = n_0 + 1$ and let $c \geq c_0$

$$\begin{aligned} |(\mathrm{GM}) \textstyle\int_{[1,c[\times[1,c[} f - A| &\leq |S_{c^*-1,c^*-1} - A| + |\sum_{i=1}^{c^*-1} a_{ic^*} + a_{c^*i}||c - c^*| \\ &\leq \varepsilon + |S_{c^*,c^*} - S_{c^*-1,c^*-1}| \\ &\leq 3\varepsilon, \text{ by}(8). \end{aligned}$$

So, we have

$$\lim_{c\to+\infty} (\mathrm{GM}) \int_{[1,c[\times[1,c[} f = A,$$

and the result follows by Theorem 2.

REFERENCES

[BU] B.Z.BUCZOLICH, Real Anal. Exchange 13(1987-88),71-75.

[ASH] J.Marshall ASH, Studies in Harmonic Analysis, Math. Assoc. of America, Washington, 76-96.

[MAW1] J.MAWHIN, Introduction á l'Analyse, Cabay(2'eme èdition), Louvain-la-Neuve, Belgique (1981).

[MAW2] J.MAWHIN, Czech. Math. J.31(1981),pp.614-632.

[MKH] A.MKHALFI, Bull. Soc. Math. Belgique 40B(1988),pp.111-130.

Université de Louvain
Institut de Mathématique
B-1348 Louvain-la-Neuve
Belgium

Integration in infinite-dimensional spaces

Ralph Henstock

Abstract. *Three exact new proofs are given of vital results. The division space integral over infinite-dimensional product spaces can be defined using a bare minimum of conditions (Theorem 1). For certain absolute and non-absolute conditions of integrability, a real functional is constant almost everywhere if cylindrical of every finite order (Theorem 2). If the integration is absolute, the integral's value is in a sense the limit almost everywhere of the integrals over some sequences of finite-dimensional sets (Theorem 4).*

This paper was written during the term of a two-year Leverhulme Trust Emeritus Fellowship award for the study of integration theory.

In countable Cartesian product spaces (sequence spaces) a result of Jessen[13] holds, so that the integral is the limit of the corresponding integral over the first n terms of the sequences. The origins of integration in uncountable Cartesian product spaces (function spaces) are in Einstein[2] and Smoluchowski[15], examples being given in Chandrasekhar[1]. Two integrals vital for theoretical physicists and chemists are in Wiener[16,17] and Feynman[3]. These integrals may be given mathematical expresssion by the integral of this paper; other expressions used are usually far too complicated. Again we can sometimes prove that the integral is the limit of integrals over spaces of finite dimensions. Muldowney[14] gives a good account of much of the theory.

The integral is a path integral using paths parametrised by points b of a set B, usually an interval $[a,+\infty)$, and we use generalized intervals that are sets of certain paths of the system. For example, if the paths are given by real-valued functions $x : B \to \mathbb{R}$ we can use the set of those paths that, when $b = b_j$, lie in the range

$$u^*(b_j) = u(j) \leq x(b_j) < v(j) = v^*(b_j), u(j) < v(j)$$

$$1 \leq j \leq n, \quad a < b_1 < \ldots < b_n.$$

A function of such generalized intervals I, is the integral

(1) $P(I) \equiv \int_{u(1)}^{v(1)} \cdots \int_{u(n)}^{v(n)} p(x_1, \ldots, x_n; C) dx_1 \ldots dx_n \; (C = (b_1, \ldots, b_n))$,

(2) $p(x_1, \ldots, x_n; C) = q(x_1; b_1 - a|x_0) q(x_2; b_2 - b_1|x_1) \ldots q(x_n; b_n - b_{n-1}|x_{n-1})$,

with q obeying a consistency condition of Smoluchowski, Chapman and Kolmogoroff, namely, that for each s in $0 < s < t$,

(3) $q(y;t|x) = \int_{-\infty}^{\infty} q(z;s|x)q(y;t-s|z)dz.$

For example, the probability measure for the path of a free spherical Brownian particle with $x(a) = 0$ is constructed taking $x_0 = 0$ and

(4) $q(y;t\,|\,x) = g(y-x;4Dt),\ g(x;t) = (\pi t)^{-\frac{1}{2}}\exp(-x^2/t).$

The positive diffusion constant D varies with the viscosity and temperature of the medium and the radius of the Brownian particle. For $D = \frac{1}{2}$, $P(I) = W(I)$, the ***Wiener measure*** of I, and properties of the integral are easy to prove. However, Feynman[3] used complex-valued q, an easy example having (4) with $D = i/4$. Such integration is very difficult to develop as the measure from P is not of generalized bounded variation, and limits under the integral sign are difficult to find. R. Johnson, I.C.I. research fellow at the Queen's University, Belfast, outlined the problem to me in 1962, and only now has a water-tight proof of a Jessen-type result been given. There are gaps in [6, 10]. The intervals at $b = b_1, \ldots, b_n$ are just like the edges of an n-dimensional rectangle, and the whole collection of paths is the definition of the Cartesian product of $(-\infty, \infty) = \mathbb{R}$, a real axis for each b.

More generally, for each $b \in B$ we have an additive division space

$$(T(b), \mathcal{T}(b), \mathcal{A}(b))$$

and from these we construct an additive division space $(T, \mathcal{T}, \mathcal{A})$ on the space of paths. To set up additive division spaces we need rather a lot of definitions. On the real line $\mathbb{R}$, the system that uses a positive function on an interval, often called the ***gauge***, and which leads to the Kurzweil-Henstock integral, satisfies all conditions pertaining to one dimension, and a flavour of the theory can be had by considering this special case.

Intervals are sets $a \le x < b$, $x < a$, and $x \ge b$, for arbitrary a, b with $a < b$, and $\delta : \mathbb{R} \to \mathbb{R}$ is used with $\delta > 0$ for all x, together with numbers M, N, such that $([a,b), x) \in \mathbf{U}$ if and only if $[a,b) \subseteq (x - \delta(x), x + \delta(x))$, with $x = a$ or $x = b$, while $((-\infty, a), -\infty) \in \mathbf{U}$, $([b, +\infty), +\infty) \in \mathbf{U}$ if and only if $a \le M, b \ge N$.

A book on division spaces will be produced in the future, to collect together known results. But in the meantime we have to rely on scattered papers such as [4] to [11], with [14].

We begin with a non-empty base space T of points (or generalized points such as paths). In T we choose some non-empty subsets called (***generalised***) ***intervals*** I, their family being $\mathcal{T}$, and a fixed non-empty family $\mathbf{V}$ of interval-point pairs (I, t) $(I \in \mathcal{T}, t \in T)$, t being an ***associated point*** of I. The integral we define depends on the kind of non-empty family $\mathcal{A}$ of some non-empty subsets $\mathbf{U}$ of $\mathbf{V}$ that we choose. We integrate over an ***elementary set*** E, i.e. an interval or a union of a finite number of mutually disjoint intervals. A subset $\mathbf{U}$ of $\mathbf{V}$ ***divides*** E if, for a finite subset $\mathcal{E}$ of $\mathbf{U}$, called a ***division*** of E from $\mathbf{U}$, the $(I,t) \in \mathcal{E}$ have mutually disjoint I with union E. The collection of these I is called a ***partition*** of E, the I being ***partial intervals*** of E from $\mathbf{U}$. A non-empty subset $\mathcal{P}$ of $\mathcal{E}$, including $\mathcal{E}$ itself, is a ***partial division*** of E from $\mathbf{U}$, while the union of the I from the $(I,t) \in \mathcal{P}$ is a ***partial set P of E that comes from $\mathcal{E}$*** and $\mathbf{U}$, and P is ***proper*** if $P \neq E$. For each elementary set E, and $\mathbf{U}.E$ the set of all $(I,t) \in \mathbf{U}$ with I a partial interval of E, let $\mathcal{A}|E$ be the set of all $\mathbf{U}.E$ dividing E with $\mathbf{U} \in \mathcal{A}$. We assume that $\mathcal{A}|E$ is not empty, saying that $\mathcal{A}$ ***divides*** E, and then that $\mathcal{A}$ is ***directed for divisions of*** E, i.e. given $\mathbf{U_1}, \mathbf{U_2} \in \mathcal{A}|E$, there is a $\mathbf{U} \in \mathcal{A}|E$ with $\mathbf{U} \subseteq \mathbf{U_1} \cap \mathbf{U_2}$. This gives a direction in $\mathcal{A}|E$, and so a Moore-Smith limit definition of the integral. The collection $(T, \mathcal{T}, \mathcal{A})$ is then called a ***division system*** for E. Given a $\mathbf{U} \in \mathcal{A}|E$, a function $h : \mathbf{U} \to I\!R$ or $h : \mathbf{U} \to \mathbf{C}$ (complex plane), and a division $\mathcal{E}$ of E from $\mathbf{U}$, $(\mathcal{E})\sum h(I,t)$ denotes the sum of $h(I,t)$ for all $(I,t) \in \mathcal{E}$. If there is a value H (in $I\!R$ or $\mathbf{C}$) such that, for each $\varepsilon > 0$ there is a $\mathbf{U} \in \mathcal{A}|E$ for which all divisions $\mathcal{E}$ of E from $\mathbf{U}$ satisfy $|(\mathcal{E})\sum h(I,t) - H| < \varepsilon$, then H is the ***integral of h over E using*** $\mathcal{A}$, written $(\mathcal{A})\int_E dh$, or $(\mathcal{A})\int_E f dk$ if $h(I,t) = f(t)k(I,t)$, omitting $(\mathcal{A})$ if it is understood. Given $\mathcal{A}, E$ and h, H is unique by the direction if H exists.

Intuitively, if the integral exists over E, it ought to exist over every partial set P of E. To this end, a ***restriction*** of $\mathbf{U}$ to P is a family $\mathbf{U}_1 \subseteq \mathbf{U}.P$. $\mathcal{A}$ has the ***restriction property*** if, for each elementary set E, each partial set P, and each $\mathbf{U} \in \mathcal{A}|E$, there is in $\mathcal{A}|P$ a restriction of $\mathbf{U}$ to P. If this holds with $(T, \mathcal{T}, \mathcal{A})$ a division system for all elementary sets E, and, given $\mathbf{U} \in \mathcal{A}|E$, if every pair P_1, P_2 of disjoint partial sets of E both come from the same division of E from $\mathcal{U}$, then we call $(T, \mathcal{T}, \mathcal{A})$ a ***division space***. (Previously, this has been called a ***non-additive division space***, but as most elementary properties of the integral can be proved for such spaces, and as a simple construction gives the rest, I have renamed the spaces.)

The further property, concerning the integrability over the union of two disjoint elementary sets, given the integrability over the separate sets, depends on another assumption concerning $\mathcal{A}$. We say that $\mathcal{A}$ is ***additive*** if, given disjoint elementary sets E_j and $\mathbf{U_j} \in \mathcal{A}|E_j$, $(j = 1,2)$, there is a $\mathbf{U} \in \mathcal{A}|E_1 \cup E_2$ with $\mathbf{U} \subseteq \mathbf{U}_1 \cup \mathbf{U}_2$.

If $\mathcal{A}$ is additive in a division space $(T, \mathcal{T}, \mathcal{A})$, we call the latter an ***additive division space*** (previously called simply a ***division space***). In this space we can prove that the property concerning P_1, P_2 holds from the other assumptions, and so is redundant here.

For E an elementary set and $\mathbf{U} \subseteq \mathbf{V}$, let $E^*.\mathbf{U}$ be the set of all t with $(I,t) \in \mathbf{U}.E$ for some I, and let the ***star-set*** E^* be the intersection of $E^*.\mathbf{U}$ for all $\mathbf{U} \in \mathcal{A}|E$. If, for every elementary set E, there is a $\mathbf{U}_E \in \mathcal{A}|E$ such that every $\mathbf{U} \in \mathcal{A}|E$ with $\mathbf{U} \subseteq \mathbf{U}_E$ has $E^*.\mathbf{U} = E^*$, we say that $(T, \mathcal{T}, \mathcal{A})$ is ***stable.***

For $X \subseteq T$ and $\mathbf{U} \subseteq \mathbf{V}$ we define

$$\mathbf{U}[X] = \{(I,t) : (I,t) \in \mathbf{U},\ t \in X\}.$$

$(T, \mathcal{T}, \mathcal{A})$ is ***fully decomposable*** (respectively, ***decomposable***) if, to every elementary set E, every family (respectively, countable family) $\mathcal{X}$ of mutually disjoint subsets $X \subseteq T$, and every function $\mathbf{U}(\cdot) : \mathcal{X} \to \mathcal{A}|E$, there is a $\mathbf{U} \in \mathcal{A}|E$ with

(5) $\mathbf{U}[X] \subseteq \mathbf{U}(X)[X]$ $(X \in \mathcal{X})$.

There is no need for the union of the $X \in \mathcal{X}$ to be T. If, for the given $\mathcal{X}$, equality occurs in (5) for all $X \in \mathcal{X}$, we call $\mathbf{U}$ the ***diagonal*** of the $\mathbf{U}(X)$.

Ordinary Riemann integration in one dimension is given by a division space without decomposability properties, and does not have the Arzelà-Lebesgue limit properties; these are given by decomposability. Full decomposability is also needed at times.

For an elementary set E and a proper partial set P we say that an additive division space $(T, \mathcal{T}, \mathcal{A})$ is ***strongly compatible*** with P in E if there is a $\mathbf{U}_P \in \mathcal{A}|E$ such that $(I,t) \in \mathbf{U}_P$ and $t \notin P^*$ imply $I \subseteq E \setminus P$ and $I^* \subseteq E^* \setminus P^*$.

Lemma 1. Let $(T, \mathcal{T}, \mathcal{A})$ be a stable, additive division space strongly compatible with P in E for all elementary sets E and all proper partial sets P. Let P, Q be proper partial sets of E with union E. Then

(6) $E^* = P^* \cup Q^*$.

(7) Further, the family consisting of the empty set, E^*, and $E^* \setminus P^*$ for all proper partial sets P of E, is a base for a topology (the intrinsic topology $\mathcal{G}_I$) for E^*

(8) and if also $(T, \mathcal{T}, \mathcal{A})$ is fully decomposable, E^* is compact for $\mathcal{G}_I$.

Proof. $E = P \cup Q$ in (6) so

$$E \supseteq P,\ \mathbf{U}.E \supseteq \mathbf{U}.P,\ E^*.\mathbf{U} \supseteq P^*.\mathbf{U} \supseteq P^*,\ E^* \supseteq P^*,\ E^* \supseteq P^* \cup Q^*.$$

To prove the opposite inequality, for the $\mathbf{U}_P$, $\mathbf{U}_Q$ of strong compatibility let $(I,t) \in \mathbf{U} \subseteq \mathbf{U}_P \cap \mathbf{U}_Q$, $t \notin P^*.\mathbf{U} = P^*$. Then $I \subseteq Q$ and

$$t \in Q^*.\mathbf{U} = Q^*,\ E^* \subseteq E^*.\mathbf{U} \subseteq P^*.\mathbf{U} \cup Q^*.\mathbf{U} = P^* \cup Q^*.$$

Hence (6). Finite intersections of sets $\setminus P^*$ are complements of finite unions of P^*, and so, by (6), are complements of some P^*. So $\mathcal{G}_I$ is the set of the empty set, E^*, and arbitrary unions of the $\setminus P^*$, giving (7). Thus in (8) we need only consider covers of E^* that are families of $\setminus P^*$ for various proper partial sets P of E. Then each $t \in E^*$ lies in at least one of the $\setminus P^*$, say $\setminus P(t)^*$. Take

$$\mathbf{U}[\mathrm{sing}(t)] \subseteq \mathbf{U}_{P(t)}\ (t \in E^*),$$

possible by full decomposability, the $\mathbf{U}_{P(t)}$ being the sets for strong compatibility. Then, for a division $\mathcal{E}$ of E from $\mathbf{U}$, and $(I,t) \in \mathcal{E}$, we have by (6)

$$(I,t) \in \mathbf{U}_{P(t)},\ t \notin P(t)^*,\ I^* \subseteq E^* \setminus P(t)^*,\ E^* = \cup_{\mathcal{E}} I^* \subseteq \cup_{\mathcal{E}} E^* \setminus P(t)^*,$$

a finite number of members of $\mathcal{G}_I$ from the cover, covers E^*, and E^* is compact in the intrinsic topology.

For Cartesian products we note that [12] gives the theory of the n - dimensional case of the Kurzweil-Henstock integral. Here we take a finite set C of more than one value c, writing x_c for the co-ordinate variable corresponding to c. Let T^C be the Cartesian product of T^c for all $c \in C$, points of T^C being written as vectors $\mathbf{x}_C$ with components x_c, $c \in C$. Let $\otimes$ denote Cartesian product. If $\mathcal{T}^c$ is the family of c - intervals $I^c \subseteq T^c$, and $\mathbf{V}^c$ the corresponding family of (I^c, x_c), for $c \in C$, let $\mathcal{T}^C, \mathbf{V}^C$ be the respective families of all $I^C \equiv \bigotimes_C I^c$ and all $(I^C, \mathbf{x}_C)$, written $\bigotimes_C (I^c, x_c)$, for all $I^c \in \mathcal{T}^c$ and all $(I^c, x^c) \in \mathbf{V}^c (c \in C)$. Let E^c be an elementary set in T^c and, for each $\mathbf{x}_C \in T^C$, let $\mathbf{U}^c(x_C) \in \mathcal{A}^c | E^c\ (c \in C)$. Let $\mathbf{U}^C$ be the family of all $\bigotimes_C (I^c, x_c)$ with I^c a partial interval of E^c and $(I^c, x_c) \in \mathbf{U}^c(\mathbf{x}_C)\ (c \in C)$, so that $\mathbf{U}^C$ is a function of the $\mathbf{U}^c(\mathbf{x}_C)$. Let $\mathcal{A}^C$ be the family of finite unions of such $\mathbf{U}^C$ for all finite unions of disjoint products $\bigotimes_C E_c$. Then $(T^C, \mathcal{T}^C, \mathcal{A}^C)$ is the ***product division space*** of the additive division spaces $(T^c, \mathcal{T}^c, \mathcal{A}^c)\ (c \in C)$, and, by construction,

(9) $(\bigotimes_C E^c)^* = \bigotimes_C (E^c)^*$.

This gives Fubini and Tonelli properties; see [6] for the former.

Next we look at the case of an infinity of Cartesian products, using an index set B that is infinite, countable or uncountable. For each $b \in B$ suppose we are given an additive division space $(T(b), \mathcal{T}(b), \mathcal{A}(b))$, which in the special case can be the one-dimensional construction for the Kurzweil-Henstock integral. To contruct a suitable $(T, \mathcal{T}, \mathcal{A})$ with T the Cartesian product of the $T(b)\ (b \in B)$, we use the $(T^C, \mathcal{T}^C, \mathcal{A}^C)$ just constructed. For each $C \subseteq B$ let $\prod(X(b); C)$ be the Cartesian product of $X(b)$ where

$$X(b) \subseteq T(b)\ (b \in C),\ X(b) = T(b)\ (b \in B \setminus C).$$

Let $X^C = X(b)^C = \bigotimes_C X(b)$, $T^C = \bigotimes_C T(b)$. In this notation, for finite C (9) becomes $E^{C*} = E^{*C}$. We take the intervals of $\mathcal{T}$ to be $I = \prod(I(b); C)$ for all ***finite*** $C \subseteq B$ and all $I(b) \in \mathcal{T}(b)(b \in C)$, while $\mathbf{V}^C, \mathbf{V}$ are the Cartesian products of $(I(b), f(b))$ for all $b \in C$, all $b \in B$, respectively, where $f : B \to K$ replaces the vector $\mathbf{x}_C$. We write $f^S : S \to K$ when $S \subseteq B$ and $f^S(b) = f(b)\ (b \in S)$. For simplicity,

(10) $T(b)$ is an elementary set relative to $\mathcal{T}(b)\ (b \in B)$.

With each constructed $\mathbf{U}$, each elementary set $E \subseteq T$, each $f \in T$, and each finite $C \subseteq B$, we associate a finite set $C_1(\mathbf{U}, f) \subseteq B$ and a $\mathbf{U}_1(C, \mathbf{U}, f) \in \mathcal{A}^C$. Then $\mathbf{U}$ is the family of all (I, f)

(11) with $I = \prod(I(b); C) \subseteq E$, C finite, $C_1(U, f) \subseteq C \subseteq B$;

(12) $(I(b), f(b))^C \in \mathbf{U}_1(C; \mathbf{U}; f)$;

(13) $f(b) \in T(b)^* \ (b \in B \setminus C)$.

For $\mathcal{A}$ the family of all such $\mathbf{U}$ for all elementary sets $E \subseteq T$, $(T, \mathcal{T}, \mathcal{A})$ is the ***product division space*** of $(T(b), \mathcal{T}(b), \mathcal{A}(b))$ $(b \in B)$. The $b \in B$ can be re-arranged without affecting the construction. B might not be sensibly ordered. By (9) and [6], p.334, Theorem 7,

(14) $I^* = I^{B*} = I^{*B}$.

Theorem 1. Let $J = \prod(J(b); C) \in \mathcal{T}$. For each $b \in B$ let a partition $\mathcal{P}_n(b)$ of $T(b)$ exist independent of any $\mathbf{U}$, such that

(15) if $I(b) \in \mathcal{P}_1(b)$ then either $I(b) \subseteq J(b)$ or $I(b) \cap J(b)$ is empty;

(16) $\mathcal{P}_{n+1}(b)$ is a refinement of $\mathcal{P}_n(b)$

i.e if $(I(b), f(b)) \in \mathcal{P}_{n+1}(b)$, there is a $(I_1(b), g(b)) \in \mathcal{P}_n(b)$ with $I(b) \subseteq I_1(b) (n = 1, 2, \ldots; b \in B)$;

(17) for each $\mathbf{U} \in \mathcal{A}$, each $f \in T$ and each finite set C^* in $C_1(\mathbf{U}, f) \subseteq C^* \subseteq B$, there is an integer N depending on $\mathbf{U}, f, \mathbf{U}_1(C^*; \mathbf{U}; f)$ and so on $\mathbf{U}, f, C^*$, such that if $n \geq N$, $I(b) \in \mathcal{P}_n(b)$, $I(b) \subseteq J(b)$, $f(b) \in I(b)^*$ then $I(b)$ is an interval $L(b)$ or a finite union of disjoint intervals $L(b)$ $(b \in C^*)$ with $(L(b), f(b))^{C^*} \in \mathbf{U}_1(C^*; \mathbf{U}; f)$;

(18) let $(T(b), \mathcal{T}(b), \mathcal{A}(b))$ be strongly compatible with every partial set of $T(b)$ $(b \in B)$.

Then $\mathbf{U}$ divides J.

(The sequence of partitions is an abstraction of continued bisection.)

Proof of theorem. Supposing J not divided by $\mathbf{U}$, we use:

(19) for $\mathcal{P}$ a partition of an elementary set E not divided by $\mathbf{U}$, an $I \in \mathcal{P}$ is not divided.

(Otherwise the union of divisions from $\mathbf{U}$ of the $I \in \mathcal{P}$ would be a division of E.) For finite C^* in $C \subseteq C^* \subseteq B$, fixed n, and all $I(b) \in \mathcal{P}_n(b)$ $(b \in C^*)$ the $\prod(I(b); C^*)$ form a partition $\mathcal{P}_n(C^*)$ of T. By (15), (16), $\mathcal{P}_{n+1}(C^*)$ refines $\mathcal{P}_n(C^*)$ and each $I = I^{C^*} \times T^{B \setminus C^*}$ with $I^{C^*} \in \mathcal{P}_n(C^*)$ has $I \subseteq J$ or $I \cap J$ empty. By (19) with $E = J$, there is an $I \in \mathcal{P}_1(C^*)$ not divided by $\mathbf{U}$, with $I \subseteq J$. If, for some $n, I \in \mathcal{P}_n(C^*)$ not divided by $\mathbf{U}$, with $I \subseteq J$, there is, by (19), an $E \in \mathcal{P}_{n+1}(C^*)$ not divided by $\mathbf{U}$, with $E \subseteq I \subseteq J$. Thus, by mathematical induction, the set $\mathcal{Q}_n(C^*)$ of all $I \in \mathcal{P}_n(C^*)$ not divided by $\mathbf{U}$, is not empty, and, for a finite number of n, C^*, the intersection of the $\mathcal{Q}_n(C^*)$ is a non-empty elementary set. By (6) the corresponding finite number of $\mathcal{Q}_n(C^*)^*$ has a non-empty intersection. By (18) and Lemma 1 (8), $T(b)$ is compact in the intrinsic topology for each $b \in B$. By Tychonoff's theorem on Cartesian products, T is compact in the intrinsic topology and for all n, C^*, the intersection of the $\mathcal{Q}_n(C^*)^*$ is not empty, and so contains g, say. Thus there is a $C_1(\mathbf{U}, g)$, (17) applies, and we have a contradiction. Hence the theorem.

Using this result it is straightforward to prove that $(T, \mathcal{T}, \mathcal{A})$ is a fully decomposable additive division space with the Fubini property relative to V, W, where B is separated into non-empty sets H and $B \setminus H$, and V, W are the respective Cartesian products of $T(b)$ for $b \in H$ and $b \in B \setminus H$. As we can integrate over V first or W first, if we are only given that the integral over T exists, the most general function for Fubini's theorem is of the type $g(f^H, f^{B\setminus H})h(I^H, f^H)k(I^{B\setminus H}, f^{B\setminus H})$, for every H. If $H = C$, finite, and $I = \prod(I(b); C)$ let $C(I)$ be the smallest subset of B with $I = \prod(I(b); C(I))$. Then we need $h(I^H, f^H)$ to be

$$(20)\quad m(C(I); I^{C(I)}; f^{C(I)}),\ m(c; I^C; f^c) = \textstyle\prod_{b\in C} m^b(I(b), f(b)),$$

$$\int_{T(b)} dm^b(I(b), f(b)) = 1,$$

the Smoluchowski-Chapman-Kolmogoroff condition. As $I^{B\setminus C(I)} = T^{B\setminus C(I)}$, k can be absorbed into g, the most general function being of the form, with (20),

$$(21)\quad g(C(I); f)m(C(I); I^{C(I)}; f^{C(I)}).$$

A functional $F : T \to I\!R$ is ***cylindrical of order*** $C \subseteq B$, if $F(f) = F(f_1)$ when $f^{B\setminus C} = f_1^{B\setminus C}$. If this holds for every finite $C \subseteq B$, F is ***cylindrical of every finite order***.

Theorem 2. In Theorem 1, with (20),

(22) a real functional F on T that is cylindrical of every finite order, with Fm integrable, is constant m-almost everywhere.

(23) A functional $F : T \to \mathbb{R}$, cylindrical of every finite order, with χm integrable for the indicator χ of each set

$$F^{-1}([p/q,(p+1)/q))\ (p = 0, \pm 1, \pm 2, \ldots; q = 1, 2, \ldots),$$

is constant m-almost everywhere.

Note that the integrability of χm is equivalent to the m-measurability of F.

Proof. For G, M the integrals of Fm, m, respectively, and $I = I^C \times T^{B\setminus C}$, Fubini's theorem and (20) give

$$G(I) = M(I) \int_{T^{B\setminus C}} F dm(B \setminus C; I^{B\setminus C}, f^{B\setminus C}),$$

$$M(T) = 1, \quad G(T) = \int_{T^{B\setminus C}} F dm(B \setminus C; I^{B\setminus C}, f^{B\setminus C}),$$

(24) $G(I) = M(I)G(T)$.

As F is cylindrical of every finite order, C can be any finite subset of B, and by (24), $G(I) - G(T)M(I)$ has variation zero. By variational equivalence, $(F(f) - G(T))m$ has variation zero and $F(f) - G(T) = 0$ m-almost everywhere, giving (22). For (23), (22) gives $\chi = 0$ m-almost everywhere or $\chi = 1$ m-almost everywhere, with $M(T) = 1$. Thus for fixed q, there is a $p = p(q)$ with $\setminus F^{-1}([p/q,(p+1)/q))$ of m-variation zero, and

$$\bigcup_{q=1}^{\infty} \setminus F^{-1}([p(q)/q,(p(q)+1)/q)) =$$

$$\setminus \bigcap_{q=1}^{\infty} F^{-1}([p(q)/q,(p(q)+1)/q)) =$$

$$\setminus F^{-1}(\bigcap_{q=1}^{\infty} [p(q)/q,(p(q)+1)/q))$$

has m-variation zero, and the last intersection contains a point r alone, so that $F = r$ m-almost everywhere.

Theorem 3. For the function of (21) with (20), and the conditions of Theorem 1 with T instead of J, let g, m be real-valued and let

$$G(E) \equiv \int_E d\{g(C(I); f)m(C(I); I^{C(I)}, f^{C(I)})\}$$

exist when $E = T$, and so for arbitrary elementary sets E in T. Then the integral

$$(25) \quad G(B \setminus C; J, f) \equiv \int_J d\{g(C(I); f)m(C(I); I^{C(I)}, f^{C(I)})\} \ (J \subseteq T^C)$$

exists for $J = T^C$ and for f $m(B \setminus C; I^{B \setminus C}, f^{B \setminus C})$-almost everywhere in $T^{B \setminus C}$, where C is finite, $C \subseteq B$, and if we put 0 for $G(B \setminus C; T^C, f)$ where it is otherwise undefined,

$$(26) \quad \int_{T^{B \setminus C}} G(B \setminus C; T^C, f) dm(B \setminus C; I^{B \setminus C}, f^{B \setminus C}) = G(T).$$

Proof. Use Fubini's theorem [6].

Theorem 4. For (10) and g real-valued, $m \geq 0$, with gm and $|gm|$ integrable over T, then, for f m-almost everywhere in T^*, $\lim^* G(B \setminus C; T, f) = G(T)$, in a sense to be explained later.

Proof. Let $\mathbf{U}_j \in \mathcal{A}$ be such that every division $\mathcal{E}$ over T from $\mathbf{U}_j$ satisfies

$$(27) \quad |(\mathcal{E}) \textstyle\sum gm - G(T)| < 1/j.$$

By direction in $\mathcal{A}$ we take the $\mathbf{U}_j$ monotone decreasing in j. By mathematical induction we choose a sequence $(\mathcal{E}_j)$ of divisions of T such that $\mathcal{E}_j$ comes from $\mathbf{U}_j$, $\mathcal{E}_{j+1}$ refining $\mathcal{E}_j$. Let C_j be the finite set of all $b \in B$ for which at least one $(I, f) \in \mathcal{E}_j$ has $b \in C(I)$. By the refinement property (C_j) is monotone increasing, the union B^* being at most countable. For fixed j the divisions $\mathcal{E}_j^*$ of T from $\mathbf{U}_j$ with $C(I) \subseteq B^*$ for all $(I, f) \in \mathcal{E}_j^*$, restricting the I but not the f, include $\mathcal{E}_j$ and are a subset of all divisions of T from $\mathbf{U}_j$, and so satisfy (27), and the corresponding restricted integral has the same value $G(T)$. Writing $(\mathcal{A}^*)$ in front of this integral, we use Jessen's kind of proof in this more general case. Let

$$B^* = (b_j), \ C^n \equiv \{b_1, \ldots, b_n\}, \ G_n(J, f) \equiv G(B \setminus C^n; J, f),$$

$$X_N(L) \equiv \{f \in T; \sup_{1 \leq n \leq N} G_n(T, f) > L\},$$

$$Y_n \equiv \{f \in T : G_n(T, f) > L\},$$

$$Z_N = Y_N, \ Z_n \equiv Y_n \setminus (Y_N \cup \ldots \cup Y_{n+1}) \ (1 \leq n < N), \ V(n) \equiv T(B \setminus C^n).$$

Each Y_j $(j \geq n)$ is cylindrical of order C^n, so that the same is true for the disjoint Z_n with $Z_n \subseteq Y_n$ and by the usual proofs, they are m-measurable $(n = 1, \ldots, N)$. Using Theorem 3,

$$\int_T \chi(Z_n; f) dm = \int_{V(n)} \chi(Z_n; f) G_n(T, f) dm$$

$$\geq L \int_{V(n)} \chi(Z_n; f) dm = L \int_T \chi(Z_n; f) dm,$$

$$X_N(L) = Y_N \cup Y_{N-1} \cup \ldots \cup Y_1 = Z_N \cup Z_{N-1} \cup \ldots \cup Z_1,$$

and by addition for $n = 1, \ldots, N$ we have

(28) $\int_T \chi(X_N(L); f) g dm \geq L \int_T \chi(X_N; f) dm.$

If $X(L), X^+(L)$ are the respective sets of $f \in T$ for which

$$\sup_n G_n(T, f) > L, \ \limsup_{n\to\infty} G_n(T, f) > L, \ X^+(L) \subseteq X(L),$$

and $X_N(L)$ is monotone increasing to $X(L)$ as $N \to \infty$. gm and $|gm|$ being integrable, the integral of gm is m-AC* so that we can let $N \to \infty$ in (28) to obtain

(29) $\int_T \chi(X(L); f) g dm \geq L \int_T \chi(X(L); f) dm,$

true for all real L. We now replace $\mathcal{A}$ by $\mathcal{A}^*$, B by B^*. When the $\mathcal{A}$-integral exists, so does the $\mathcal{A}^*$-integral, with the same value, so that from (29),

(30) $(\mathcal{A}^*) \int_T \chi(X(L); f) g dm \geq L.(\mathcal{A})^* \int_T \chi(X(L); f) dm.$

From the last, $X(L)$ is m-measurable using $\mathcal{A}^*$, and by the usual arguments, so is $X^+(L)$. It is cylindrical of order B^*, so that, by Theorem 2 for B^*, $\mathcal{A}^*$, and some constant H,

$$\limsup_{n\to\infty} G_n(T, f) = H$$

m-almost everywhere using $\mathcal{A}^*$. Taking $L < H, \backslash X^+(L)$, and so $\backslash X(L)$, have m-variation zero using $\mathcal{A}^*$, and (30) becomes

$$G(T) = (\mathcal{A}^*) \int_T g dm \geq L.(\mathcal{A}^*) \int_T dm = L, \ G(T) \geq H = \limsup_{n\to\infty} G_n(T, f)$$

m-almost everywhere using $\mathcal{A}^*$. Replacing g in the proof by $-g$, we have

$$G(T) \leq \liminf_{n\to\infty} G_n(T, f)$$

m-almost everywhere, using $\mathcal{A}^*$, so that with $\mathcal{A}^*$, and m-almost everywhere,

(31) $\lim_{n\to\infty} G_n(T, f) = G(T).$

Replacing B^* by a countable B^{**} in $B^* \subseteq B^{**} \subseteq B$, (31) follows again for a suitable $\mathcal{A}^{**}$. This is the sense of the statement involving $\lim^*$.

References

1. S.Chandrasekhar, 'Stochastic problems in physics and astronomy', ***Reviews of Modern Physics*** 15 (1943) 1-89, ***Math. Rev.*** 4-248.

2. A. Einstein, 'Zur Theorie der Brownschen Bewegung', ***Ann. d. Physik*** 19 (1906) 371-381.

3. R. P. Feynman, 'Space-time approach to non-relativistic quantum mechanics', ***Reviews of Modern Physics*** 20(1948) 367-387, ***Math. Rev.*** 10-224.

4. R. Henstock, ***Linear Analysis*** (Butterworths, London, 1968), ***Math. Rev.*** 54 #7725.

5. R. Henstock, 'Generalized integrals of vector-valued functions', ***Proc. London Math. Soc.*** (3) 19(1969) 509-536, ***Math. Rev.*** 40 #4420.

6. R. Henstock, 'Integration in product spaces, including Wiener and Feynman integration', ***Proc. London Math. Soc.*** (3) 27 (1973) 317-344, ***Math. Rev.*** 49 #9145.

7. R. Henstock, 'Additivity and the Lebesgue limit theorems', ***the Greek Mathematical Society C. Carathéodory Symposium*** Sept. 3-7, 1973, pp. 223-241 (published 1974), ***Math. Rev.*** 57 #6355.

8. R. Henstock, 'Integration, variation and differentiation in division spaces', ***Proc. Royal Irish Academy***, Series A, 78 (1978) No.10, 69-85, ***Math. Rev.*** 80d:26011.

9. R. Henstock, 'Generalized Riemann integration and an intrinsic topology', ***Canadian Journal of Mathematics*** 32(1980) 395-413, ***Math. Rev.*** 82b:26010.

10. R. Henstock, 'Division spaces, vector-valued functions and backwards martingales', ***Proc. Royal Irish Academy***, series A, 80(1980) No. 2, 217-232, ***Math. Rev.*** 82i:60091.

11. R. Henstock, 'Density integration and Walsh functions', ***Bull. Malaysian Math. Soc.*** (2)5(1982) 1-19, ***Math. Rev.*** 84i:26010.

12. R. Henstock ***Lectures on the Theory of Integration*** (World Scientific, Singapore, 1988).

13. B. Jessen, 'The theory of integration in a space of an infinite number of dimensions', ***Acta Math.*** 63(1934) 249-323, ***Zblt.*** 10.200

14. P. Muldowney, ***A general theory of integration in function spaces*** (Longman, Pitman Research Notes in Math. 153, 1987).

15. M Smoluchowski, ***Ann. d. Physik*** 48 (1915) 1103.

16. N. Wiener, 'Differential space', ***J. Math. Phys.*** 2(1923) 132-174.

17. N. Wiener, 'Generalized harmonic analysis', ***Acta Math.*** 55 (1930) 117-258.

University of Ulster

Coleraine

N. Ireland

The PU-integral: its definition and some basic properties

Jaroslav Kurzweil and Jiří Jarník

INTRODUCTION

Let $f : \mathbb{R}^n \to \mathbb{R}$ have a compact support. The PU-integral of f (over $\mathbb{R}^n$) is defined as a limit of Riemannian sums $\sum_i f(t^i) \int \vartheta_i dx$ and denoted by $(\mathrm{PU})\int f dx$; here $t^i \in$ supp f and $\{\vartheta_i : i \in \mathcal{T}\}$ is a partition of unity on supp f. ($\mathcal{T}$ is a finite set).

The PU-integral is an extension of the Lebesgue integral. It is nonabsolutely convergent, admits transformation by $C^{(1)}$-diffeomorphisms so that it can be lifted to manifolds, and makes it possible to give a general formulation of the Stokes' Theorem.

A survey of the main features of the PU-integral was presented at the Twelfth Summer Symposium in Real Analysis, Coleraine 1988, and has been published in Real Analysis Exchange.

In this paper we verify the correctness of the definition of the PU-integral, and then we prove

(i) every continuously differentiable function is a multiplier of any PU-integrable function,

(ii) an analogue of the Saks-Henstock Lemma holds.

The key role in the proof of both results is played by the Lemma on Multiplication of Systems; its role can be compared to the role of the partitioning property in Henstock's theory. The correctness of the definition of the PU-integral reduces to the existence of some special partitions of unity which in turn follows in a standard manner from a covering lemma; this lemma guarantees the existence of an arbitrarily fine cover of a compact set by tagged sets which fulfil a regularity (with respect to the tag) condition and whose interiors are pairwise disjoint.

1 Notation and preliminaries

Let $\mathbb{R}^n, n = 1, 2, \ldots$ be the Euclidean space with the norm denoted by $||\cdot||$. The symbols $B(x, \eta)$ and $\bar{B}(x, \eta)$ stand respectively for the open and closed ball with center $x \in \mathbb{R}^n$ and $\eta > 0$. By $C^{(1)}$ we denote the class of functions continuous on $\mathbb{R}^n$ together with their first derivatives. The support of a function f is denoted by supp f, its derivative by Df.

Any family

$$\theta = \{(t^i, \vartheta_i); i \in \mathcal{T}\} \tag{1}$$

will be called a system provided

$$\mathcal{T} \text{ is a finite set;} \tag{2}$$

$$t^i \in \mathbb{R}^n \text{ for } i \in \mathcal{T}; \tag{3}$$

$$\vartheta_i : \mathbb{R}^n \to [0,1] \text{ for } i \in \mathcal{T} \text{ has a compact support and } \vartheta_i \in C^{(1)} \tag{4}$$

$$0 \le \sum_{i \in \mathcal{T}} \vartheta_i(x) \le 1 \text{ for } x \in \mathbb{R}^n. \tag{5}$$

We denote (for $i \in \mathcal{T}$)

$$\rho_i = \rho_i(\theta) = \sup\{||x - t^i||;\ x \in \ \text{supp}\ \vartheta_i\}.$$

Let $Q \subset \mathbb{R}^n$ be compact. The system θ from (1) is called a PU-cover of Q (PU standing for "partition of unity") if

$$t^i \in Q \text{ for } i \in \mathcal{T}; \tag{6}$$

$$\{x \in \mathbb{R}^n;\ \sum_{i \in \mathcal{T}} \vartheta_i(x) = 1\} \supset Q. \tag{7}$$

Any function $\delta : Q \to (0,1)$ will be called a gauge on Q. (Note that the restriction of the range of δ to $(0,1)$ is not essential. The same results are obtained with $\delta : Q \to (0,\infty)$.)

Let θ be the system defined by (1) and Q a set such that $t^i \in Q$, $i \in \mathcal{T}$. If δ is a gauge on Q, then the system θ is said to be δ-fine if

$$\text{supp}\ \vartheta_i \subset B(t^i, \delta(t^i))$$

or, equivalently,

$$\rho_i < \delta(t^i).$$

Let $K \ge 1$, $L \ge 1$ be constants, $\zeta : Q \times (0,1] \to (0,\infty)$ a function satisfying

$$\zeta(t,\sigma) \nearrow \infty,\ \sigma\zeta(t,\sigma) \searrow 0 \text{ for } \sigma \searrow 0 \tag{8}$$

for $t \in Q$.

The system (1) is said to be K, L, ζ-regular if

$$||D\vartheta_i(x)|| \le \zeta(t^i, \rho_i)\vartheta_i(t^i) \text{ for } x \in B(t^i, \frac{\rho_i}{K}); \tag{9}$$

if $||D\vartheta_i(x)|| > \zeta(t^i, \rho_i)\vartheta_i(t^i)$ for some $x \in \mathbb{R}^n$, then

$$D\vartheta_i(x).(t^i - x) \geq \frac{1}{L}||D\vartheta_i(x)||.||t^i - x||. \tag{10}$$

Condition (10) can be interpreted geometrically as follows. Denote by $\alpha_i(x)$ the angle of the vectors $D\vartheta_i(x)$ and $t^i - x$. If $||D\vartheta_i(x)||$, i.e. the norm of the gradient of ϑ_i at x, is large (does not fulfil the inequality from (9)) then

$$\cos \alpha_i(x) \geq \frac{1}{L} > 0.$$

In other words, in this case the angle of the vectors $D\vartheta_i(x)$ and $t^i - x$ must be strictly separated from $\frac{\pi}{2}$.

2 Existence and properties of PU-covers

Theorem 2.1. There exist constants $K_0 \geq 1$, $L_0 \geq 1$ such that for every compact set $Q \subset I\!R^n$ and every gauge δ on Q there is a δ-fine PU-cover θ of Q satisfying

$$\theta = \{(t^i, \vartheta_i);\ i \in \mathcal{T}\}; \tag{11}$$

$$D\vartheta_i(x) = 0 \text{ for } x \in B(t^i, \frac{\rho_i}{K_0}); \tag{12}$$

$$D\vartheta_i(x).(t^i - x) \geq \frac{1}{L_0}||D\vartheta_i(x)||.||t^i - x|| \text{ for } x \in I\!R^n. \tag{13}$$

The cover θ from the theorem will be called a standard cover of Q. Evidently, it is a K, L, ζ-regular PU-cover of Q for any ζ satisfying (8) and for any K, L; $K \geq K_0, L \geq L_0$. Notice that the constants K_0, L_0 depend only on the dimension n.

To prove the existence of standard covers, we first construct a suitable set cover of Q (Theorem 2.2). Then the desired PU-cover may be obtained by regularizing the characteristic functions of the individual sets. Since the latter step is rather technical, we omit it in the present paper.

Theorem 2.2. Let $Q \subset I\!R^n$ be compact, let δ be a gauge on Q. Then there exists a family

$$\Delta = \{(t^j, H_j);\ t^j \in Q,\ H_j \subset I\!R^n \text{ compact}, j = 1, 2, \ldots, k\} \tag{14}$$

such that, denoting $\tau_j = \sup\{||x - t^j||;\ x \in H_j\}$, we have

$$0 < \tau_j \leq \delta(t^j), \tag{15}$$

$$\bar{B}(t^j, \frac{1}{2}\tau_j) \subset H_j, \tag{16}$$

$$\text{conv}\,(\{x\} \cup \bar{B}(t^j, \frac{1}{2}\tau_j)) \subset H_j \text{ for } x \in H_j,\ j = 1, 2, \ldots, k. \tag{17}$$

Moreover, the sets H_j form a nonoverlapping cover of Q, i.e.

$$\bigcup_{j=1}^{k} H_j \supset Q, \text{ Int } H_j \cap \text{ Int } H_i = \emptyset \text{ for } j \neq i. \tag{18}$$

Proof. First, we will construct two finite sequences of points $t^j \in Q$ and positive numbers σ_j, $j = 1, 2, \ldots, k$ such that

$$\sigma_1 > \sigma_2 > \ldots > \sigma_k > 0, \tag{19}$$

$$\bigcup_{j=1}^{k} B(t^j, \sigma_j) \supset Q, \tag{20}$$

$$||t^j - t^i|| \geq \sigma_i \text{ for } i < j \leq k. \tag{21}$$

The construction proceeds as follows. Fix a sequence of numbers η_l,

$$\frac{1}{4} > \eta_1 > \eta_2 > \ldots > 0, \sum_{l=1}^{\infty} \eta_l < \infty,$$

choose $t^1 \in Q$ such that

$$\delta(t^1) \geq (1 - \eta_1) \sup\{\delta(t);\ t \in Q\},$$

and set $\sigma_1 = (1 - \eta_1)\delta(t^1)$.

Assume that we have found $t^1, \ldots, t^m \in Q$, $\sigma_1, \sigma_2, \ldots, \sigma_m \in (0, \infty)$, $m \geq 1$, such that (19), (21) hold for $k = m$, and

$$Q_m = Q \setminus \bigcup_{j=1}^{m} B(t^j, \sigma_j) \neq \emptyset.$$

Choose $t^{m+1} \in Q_m$ such that

$$\delta(t^{m+1}) \geq (1 - \eta_{m+1}) \sup\{\delta(t);\ t \in Q_m\}$$

and σ_{m+1} such that

$$\sigma_m > \sigma_{m+1} \geq (1 - \eta_{m+1}) \min\{\delta(t^{m+1}), \sigma_m\}.$$

We claim that after a finite number of steps the process stops since the union of balls $B(t^j, \sigma_j)$ covers the set Q.

On the contrary, let us assume that this is not the case so that we have infinite sequences of t^j, σ_j. Since Q is compact, necessarily $\sigma_j \searrow 0$, and there is $y \in Q \setminus \cup_{j=1}^{\infty} B(t^j, \sigma_j)$. Obviously $\delta(y) > 0$, hence by definition

$$\delta(t^j) \geq (1 - \eta_j)\delta(y) \geq \frac{3}{4}\delta(y) > 0$$

for all $j = 1, 2, \ldots$. Consequently, there is l such that $\sigma_l < \frac{3}{4}\delta(y)$. For $j \geq l$ we have $\sigma_j < \frac{3}{4}\delta(y) \leq \delta(t^j)$, hence also

$$\sigma_{j+1} \geq (1 - \eta_{j+1})\sigma_j \text{ for } j \geq l$$

and

$$\sigma_{j+1} \geq (1 - \eta_{j+1})(1 - \eta_j) \ldots (1 - \eta_{l+1})\sigma_l > c > 0$$

by virtue of convergence of the series $\sum_{l=1}^{\infty} \eta_l$. This is a contradiction with $\sigma_j \searrow 0$, hence the above described process necessarily stops after a finite number, say k, of steps.

Now define $\omega_j : I\!\!R^n \to [0, \infty), j = 1, 2, \ldots, k$ by

$$\begin{aligned} \omega_j(x) &= 0 \text{ if } \|x - t^j\| \geq \sigma_j, \\ &= \sigma_j - \|x - t^j\| \text{ if } \|x - t^j\| < \sigma_j \end{aligned}$$

and set

$$H_j = \text{Cl}\{x;\ \omega_j(x) > \omega_i(x) \text{ for } i \neq j,\ i \in \{1, 2, \ldots, k\}\} = \bigcap_{i \neq j} G_{ji},$$

where

$$G_{ji} = \text{Cl}\{x;\ \omega_j(x) > \omega_i(x)\} \text{ for } i \neq j.$$

Since the sets H_j obviously satisfy (15), we have to prove (16), (17). Let $1 \leq i < j \leq k$. We will prove that

$$\bar{B}(t^j, \frac{1}{2}\tau_j) \subset G_{ji}, \tag{22}$$

$$\text{conv}\{\{x\} \cup \bar{B}(t^j, \frac{1}{2}\tau_j)\} \subset G_{ji} \text{ for } x \in G_{ji}, \tag{23}$$

$$\bar{B}(t^i, \frac{1}{2}\tau_i) \subset G_{ij}, \tag{24}$$

$$\text{conv}\{\{x\} \cup \bar{B}(t^i, \frac{1}{2}\tau_i)\} \subset G_{ij} \text{ for } x \in G_{ij}. \tag{25}$$

Then (22) - (25) together with $H_j = \cap_{i \neq j} G_{ji}$ will guarantee that (16), (17) hold.

Without loss of generality we assume that $\|t^i - t^j\| < \sigma_i + \sigma_j$. Let us denote $2a = \|t^i - t^j\|$, $2l = \sigma_i - \sigma_j$ so that $0 < 2l < 2a$, $\sigma_i \leq 2a < \sigma_i + \sigma_j$. Choose a special coordinate system such that

$$t^i = (-a, 0, \ldots, 0),\ t^j = (a, 0, \ldots, 0).$$

Further, let us set

$$\begin{aligned} A &= \{x = (x_1, \ldots, x_n);\ x_1 > 0,\ \tfrac{x_1^2}{l^2} - \tfrac{x_2^2 + \cdots + x_n^2}{a^2 - l^2} \geq 1\}, \\ C &= \text{Cl}(I\!\!R^n \setminus A) \\ &= \{x;\ x_1 \leq 0\} \cup \{x;\ x_1 > 0,\ \tfrac{x_1^2}{l^2} - \tfrac{x_2^2 + \cdots + x_n^2}{a^2 - l^2} \leq 1\}. \end{aligned}$$

Then

$$G_{ji} = A \cap \bar{B}(t^j, \sigma_j), \tag{26}$$

$$G_{ij} = C \cap \bar{B}(t^i, \sigma_i). \tag{27}$$

Since $\tau_j \leq \sigma_j$, (22) will hold if we prove

$$\bar{B}(t^j, \frac{1}{2}\sigma_j) \subset A.$$

To prove this inclusion we have to estimate the distance of t^j from the boundary of A. For simplicity, denote $v^2 = x_2^2 + \cdots + x_n^2$. We have

$$||x - t^j||^2 = (x_1 - a)^2 + v^2 = (x_1 - a)^2 + \frac{a^2 - l^2}{l^2}(x_1^2 - l^2)$$

(we substitute for v from the definition of A) and elementary calculation yields

$$||x - t^j||^2 = (x_1 \frac{a}{l} - l)^2.$$

From the definition of A we have $x_1 \geq l$, hence the minimum of the left-hand side is achieved for $x_1 = l$, being equal to $(a - l)^2$. According to (21) we have $2a \geq \sigma_i$, and using the identity $2l = \sigma_i - \sigma_j$ we conclude

$$||x - t^j||^2 \geq \frac{1}{2}\sigma_j \geq \frac{1}{2}\tau_j.$$

(22) is proved. To prove (23) it suffices to show that A is convex, which can be done by elementary calculation.

Now we have to prove (24), (25). By (21) we have $-a + \frac{1}{2}\sigma_i \leq 0$, hence

$$\bar{B}(t^i, \frac{1}{2}\tau_i) \subset \bar{B}(t^i, \frac{1}{2}\sigma_i) \subset \{x;\ x_1 \leq 0\} \subset C$$

and (24) holds.

To prove (25) let $w \in G_{ij}$. Without loss of generality we will assume that $w = (w_1, w_2, 0, \ldots, 0)$, $w_2 \geq 0$.

Denote by $z = (z_1, \ldots, z_n)$ the intersection point of the hyperspheres with centres t^i, t^j and radii σ_i, σ_j, respectively, such that $z_2 > 0, z_3 = \ldots = z_n = 0$. With each point $w = (w_1, w_2, 0, \ldots, 0) \in G_{ij}$ let us associate a point $y = (y_1, y_2, 0, \ldots, 0)$ in the following way:

- if either $w_1 \leq -a$ or $w_1 > -a$ and $w_2/(w_1 + a) \geq z_2/(z_1 + a)$ then put $y \equiv z$;
- if $w_1 > -a$ and $w_2/(w_1 + a) < z_2/(z_1 + a)$ then let y be the point satisfying

$$\frac{y_1^2}{l^2} - \frac{y_2^2}{a^2 - l^2} = 1,\ y_2 = \frac{w_2}{w_1 + a}(y_1 + a).$$

Put

$$W_y = \{u = (u_1, u_2, \dots, u_n);\ \frac{y_1(u_1 - y_1)}{l^2} - \frac{y_2(u_2 - y_2)}{a^2 - l^2} \le 0\}$$

(a halfspace whose boundary hyperplane is tangent to C at y). By elementary methods of (plane) analytical geometry we find that

$$W_y \subset C,$$

$$w \in G_{ij} \Rightarrow w \in W_y.$$

If we prove that

$$\bar{B}(t^i, \frac{\sigma_i}{2}) \subset W_y, \tag{28}$$

then evidently

$$\operatorname{conv}\{\{x\} \cup \bar{B}(t^i, \frac{\sigma_i}{2})\} \subset W_y \subset C$$

and (25) will follow immediately.

To prove (28) we will estimate the distance of the point t_i from the boundary hyperplane of W_y. We have

$$l \le y_1 \le l\frac{\sigma_i + \sigma_j}{2a},\ y_2^2 = (a^2 - l^2)\frac{y_1^2 - l^2}{l^2}, \tag{29}$$

hence the equation of the boundary hyperplane can be written in the form

$$u_1 y_1 (a^2 - l^2) - u_2 y_2 l^2 - (a^2 - l^2) l^2 = 0.$$

The distance of the point $t^i = (-a, 0, \dots, 0)$ from this hyperplane is

$$d = \frac{a y_1 (a^2 - l^2) + (a^2 - l^2) l^2}{[y_1^2 (a^2 - l^2)^2 + y_2^2 l^4]^{1/2}} = (a^2 - l^2)^{1/2} (\frac{a y_1 + l^2}{a y_1 - l^2})^{1/2}.$$

Taking into account (29), $2l = \sigma_i - \sigma_j$ and $2a \ge \sigma_i > \sigma_j$ we can estimate

$$\begin{aligned} d &\ge (a^2 - l^2)^{1/2}(1 + \tfrac{2l^2}{a y_1 - l^2})^{1/2} \\ &\ge (a^2 - l^2)^{1/2}(1 + \tfrac{2l^2}{\frac{1}{2} l(\sigma_i + \sigma_j) - l^2})^{1/2} \\ &\ge \tfrac{1}{2}[\sigma_i^2 - (\sigma_i - \sigma_j)^2]^{1/2}(1 + \tfrac{\sigma_i - \sigma_j}{\sigma_j})^{1/2} \\ &= \tfrac{1}{2}[\sigma_j(2\sigma_i - \sigma_j)\tfrac{\sigma_i}{\sigma_j}]^{1/2} \ge \tfrac{1}{2}\sigma_i, \end{aligned}$$

and (25) follows.

The next lemma is of crucial importance for the major part of the results of the present paper.

Lemma MS (on multiplication of systems). Let Q be a subset of $I\!R^n$, $K \geq 1$, $L \geq 1$ constants, δ a gauge on Q, $\zeta : Q\times(0,1] \to (0,\infty)$ a function satisfying (8). Let K_1, η be positive constants, $\eta < 1$.

Then there exists a gauge δ_1 on Q, $\delta_1(x) \leq \delta(x)$ for $x \in Q$, such that the following assertion holds:

If (11) is a PU-cover of Q which is δ_1-fine and K, L, ζ-regular, and if $\phi : \Omega \to [\eta,\ 1]$ defined on an open set $\Omega \supset Q$ is of class $C^{(1)}$ and satisfies

$$||D\phi|| \leq K_1 \text{ for } x \in \Omega,$$

then the system

$$\phi\theta = \{(t^i,\ \phi\vartheta_i);\ t_i \in T\}$$

is δ_1-fine and $K, 2L, 3\zeta$-regular.

Proof. Since evidently supp $\vartheta_i =$ supp $\phi\vartheta_i$ (notice that $\phi(x) \geq \eta > 0$ for $x \in \Omega$), the system $\phi\theta$ is obviously δ_1-fine and $\rho_i(\phi\theta) = \rho_i(\theta)$ for $i \in T$.

To prove the analogue of (9) for the system $\phi\theta$ we estimate

$$||D(\phi\vartheta_i)(x)|| \leq \phi(x)||D\vartheta_i(x)|| + \vartheta_i(x)||D\phi(x)||$$

for $x \in B(t^i,\ \frac{\rho_i}{K})$. Taking into account the inequalities

$$\vartheta_i(x) \leq \vartheta_i(t^i) + \frac{\rho_i}{K}\sup\{||D\vartheta_i(x)||;\ x \in B(t^i,\ \frac{\rho_i}{K})\},$$

$\phi(x) \leq \phi(t^i) + \rho_i K_1/K$, $\phi(x) \geq \eta$ and (9) we conclude

$$||D(\phi\vartheta_i)(x)|| \leq \phi(t^i)\vartheta_i(t^i)\zeta(t^i,\ \rho_i)(1 + \frac{2\rho_i K_1}{K\eta} + \frac{K_1}{\zeta(t^i,\ \rho_i)\eta}).$$

Consequently, the desired inequality

$$||D(\phi\vartheta_i)(x)|| \leq 3\zeta(t^i,\ \rho_i)\vartheta_i(t^i)\phi(t^i)$$

holds if δ_1 is such that

$$\frac{K_1}{\zeta(t,\ \delta_1(t))} + \frac{2\delta_1(t)K_1}{K} \leq 2\eta. \tag{30}$$

To complete the proof of Lemma MS, we have to prove: if

$$||D(\phi\vartheta_i)(x)|| > 3\zeta(t^i,\ \rho_i)\vartheta_i(t^i)\phi(t^i) \tag{31}$$

for some $x \in Q$, then

$$D(\phi\vartheta_i)(x).(t^i - x) \geq \frac{1}{2L}||D(\phi\vartheta_i)(x)||.||t^i - x||. \tag{32}$$

Let (31) hold. Then

$$\phi(x)||D\vartheta_i(x)|| \geq 3\zeta(t^i,\,\rho_i)\vartheta_i(t^i)\phi(t^i) - \vartheta_i(x)||D\phi(x)||,$$

$$||D\vartheta_i(x)|| \geq 3\zeta(t^i,\,\rho_i)\vartheta_i(t^i)\frac{\phi(t^i)}{\phi(x)} - \vartheta_i(x)\frac{K_1}{\eta}. \tag{33}$$

To estimate the ratio $\phi(t^i)/\phi(x)$ we notice that

$$\begin{aligned}\phi(x) &\leq \phi(t^i) + K_1||x - t^i|| \leq \phi(t^i) + K_1\delta_1(t^i)\\ &\leq \phi(t^i)(1 + \tfrac{K_1}{\eta}\delta_1(t^i)) \leq \tfrac{3}{2}\phi(t^i)\end{aligned}$$

provided

$$\delta_1(t) \leq \frac{\eta}{2K_1} \text{ for } t \in Q. \tag{34}$$

Similarly, we have

$$\vartheta_i(x) \leq \vartheta_i(t^i) + \rho_i\zeta(t^i,\,\rho_i)\vartheta_i(t^i).$$

Indeed, to verify this inequality we express $\vartheta_i(x)$ in the form

$$\vartheta_i(x) = \vartheta_i(t^i) + \int_0^1 D\vartheta_i(t^i + \lambda(x - t^i))d\lambda.(x - t^i)$$

and notice that if $D\vartheta_i(y).(y - t^i) > 0$ then (10) implies $||D\vartheta_i(y)|| \leq \zeta(t^i, \rho_i)\vartheta_i(t^i)$.

Assuming

$$\delta_1(t)\zeta(t,\,\delta_1(t)) \leq 1 \text{ for } t \in Q \tag{35}$$

we obtain from the inequality following (34) that

$$\vartheta_i(x) \leq 2\vartheta_i(t^i), \tag{36}$$

and inserting this together with the above estimate of $\phi(x)/\phi(t^i)$ into (33) we obtain

$$||D\vartheta_i(x)|| \geq 2\vartheta_i(t^i)(\zeta(t^i,\,\rho_i) - \frac{K_1}{\eta}).$$

Assuming further

$$\zeta(t^i,\,\delta_1(t^i)) > \frac{2K_1}{\eta} \tag{37}$$

we conclude

$$||D\vartheta_i(x)|| > \zeta(t^i,\,\rho_i)\vartheta_i(t^i). \tag{38}$$

Consequently, by (10) we have

$$D\vartheta_i(x).(t^i - x) \geq \frac{1}{L}||D\vartheta_i(x)||.||t^i - x||, \tag{39}$$

and combining our estimates we obtain

$$\begin{aligned}
D(\phi\vartheta_i)(x).(t^i - x) &= \phi(x)D\vartheta_i(x).(t^i - x) + \vartheta_i(x)D\phi(x).(t^i - x) \\
&\geq \tfrac{1}{L}\phi(x)||D\vartheta_i(x)||.||t^i - x|| - \vartheta_i(x)K_1||t^i - x|| \\
&\geq \tfrac{1}{2L}||D(\phi\vartheta_i)(x)||.||t^i - x|| - \tfrac{1}{2L}\vartheta_i(x)K_1||t^i - x|| \\
&\quad + \tfrac{1}{2L}\phi(x)\zeta_i(t^i, \rho_i)\vartheta_i(t^i)||t^i - x|| - \vartheta_i(x)K_1||t^i - x|| \\
&\geq \tfrac{1}{2L}||D(\phi\vartheta_i(x)||.||t^i - x|| \\
&\quad + \tfrac{1}{2L}||t^i - x||\vartheta_i(t^i)[\eta\zeta(t^i, \rho_i) - 2(1 + 2L)K_1]
\end{aligned}$$

(cf. (39), (38), (36)). Hence

$$D(\phi\vartheta_i)(x).(t^i - x) \geq \frac{1}{2L}||D(\phi\vartheta_i)(x)||.||t^i - x||$$

provided we assume

$$\zeta(t, \ ; \delta_1(t)) \geq 4K_1(1 + L)/\eta. \tag{40}$$

Consequently, if δ_1 is chosen so that (30), (34), (35), (37), (40) are fulfilled (which can be achieved by diminishing the values of δ_1 if necessary) then the assertion of Lemma MS holds, which completes the proof.

3 Definition of the PU-integral

Definition 3.1. Let $f : I\!R^n \to I\!R$ have a compact support, let $\gamma \in I\!R$. For any ε, K, L such that $\varepsilon > 0$, $K \geq 1$, $L \geq 1$ let there exist a gauge δ on supp f and a function ζ satisfying (8) with $Q =$ supp f such that

$$|\gamma - \sum_{i \in T} f(t^i) \int \vartheta_i dx| \leq \varepsilon$$

provided $\theta = \{(t^i, \vartheta_i);\ i \in T\}$ is any δ-fine K, L, ζ-regular PU-cover of supp f. The number γ is then called the PU-integral of f and we write $\gamma = (\mathrm{PU}) \int f dx$.

Remark 3.2. Since the standard PU-cover whose existence was established in the previous section (for any gauge δ) is K, L, ζ-regular for any $K \geq K_0$, $L \geq L_0$ and any ζ satisfying (8), our definition makes good sense.

Similarly as in [1] and [2] we give an alternative definition of integral which will be shown to be equivalent to the PU-integral from Definition 2.1.

Definition 3.3. Let $f : I\!R^n \to I\!R$, $I \subset I\!R^n$ a compact interval, supp $f \subset I$. Replace supp f by I in Definition 2.1. Then the number γ is called the PUI-integral of f, notation $\gamma = (\mathrm{PUI}) \int f dx$.

Theorem 3.4. If (PUI)$\int f dx$ exists, then (PU)$\int f dx$ exists as well and

$$(\mathrm{PUI})\int f dx = (\mathrm{PU})\int f dx.$$

Proof. Let (PUI) $\int f dx$ exist. Given $\varepsilon > 0$, $K \geq K_0$ $L \geq 2L_0$ let us fix δ and ζ so that

$$|(\mathrm{PUI})\int f dx - \sum_{j\in\mathcal{J}} f(s^j)\int \xi_j dx| \leq \varepsilon$$

for every δ-fine K, L, ζ-regular PU-cover $\Xi = \{(s^j,\, \xi_j);\ j \in \mathcal{J}\}$ of I. Let $\theta = \{(t^i,\, \vartheta_i);\ i \in \mathcal{T}\}$ be a δ-fine K, $\frac{L}{2}$, $\frac{\zeta}{3}$-regular PU-cover of supp f and let G be an open set, $B(t, 1) \subset G$ for $t \in I$. Denote

$$\vartheta(x) = \sum_{i\in\mathcal{T}} \vartheta_i(x),\ \phi(x) = 1 - \frac{1}{2}\vartheta(x),$$

set $\eta = \frac{1}{2}$, $K_1 = \frac{1}{2}\max\{||D\vartheta(x)||;\ x \in \mathbb{R}^n\}$, $Q = I$ and find δ_1 : supp $f \to (0,\, 1]$ from Lemma MS, assuming without loss of generality that

$$|f(t)|\max\{1 - \vartheta(x);\ x \in B(t, \delta_1(t))\} \leq \frac{\varepsilon}{m(G)} \tag{41}$$

for $t \in$ supp f, where m stands for the Lebesgue measure.

Let

$$\Lambda = \{(t^j,\, \vartheta_j);\ j \in \mathcal{J}\}$$

be a δ_1-fine standard PU-cover of I existing by virtue of Theorem 2.1. Λ satisfies (12), (13) and hence it is K, $\frac{L}{2}$, $\frac{\zeta}{3}$-regular.

By virtue of Lemma MS, the system

$$(1 - \frac{1}{2}\vartheta)\Lambda = \{(t^j,\, (1 - \frac{1}{2}\vartheta)\vartheta_j;\ j \in \mathcal{J}\}$$

is δ_1-fine and satisfies (9), (10) with the parameters K, L, ζ after the obvious changes of notation. Evidently, the system $(1 - \frac{1}{2}\vartheta)\Lambda \cup \frac{1}{2}\theta$ is a δ-fine PU-cover of I which is K, L, ζ-regular. (Note that $\vartheta(x) = 1$ for $x \in$ supp f; the meaning of $\frac{1}{2}\theta$ is clear.) Consequently,

$$|(\mathrm{PUI})\int f dx - \sum_{j\in\mathcal{J}} f(t^j)\int(1 - \frac{1}{2}\vartheta)\vartheta_j dx - \sum_{i\in\mathcal{T}} f(t^i)\frac{1}{2}\int \vartheta_i dx| \leq \varepsilon,$$

hence

$$\frac{1}{2}|(\mathrm{PUI})\int f dx - \sum_{i\in\mathcal{T}} f(t^i)\int \vartheta_i dx|$$

$$\leq \varepsilon + |\frac{1}{2}(\mathrm{PUI})\int f dx - \sum_{j\in\mathcal{J}} f(t^j)\int (1-\frac{1}{2}\vartheta)\vartheta_j dx|.$$

On the other hand, we have

$$|\sum_{j\in\mathcal{J}} f(t^j)\int (1-\frac{1}{2}\vartheta)\vartheta_j dx - \sum_{j\in\mathcal{J}} \frac{1}{2} f(t^j)\int \vartheta_j dx|$$

$$\leq \frac{1}{2}\sum_{j\in\mathcal{J}} |f(t^j)| \max\{1-\vartheta(x);\ x\in B(t^j,\ \delta(t^j))\}\int \vartheta_j dx$$

$$\leq \frac{\varepsilon}{2m(G)}\sum_{j\in\mathcal{J}}\int \vartheta_j dx \leq \frac{1}{2}\varepsilon$$

(cf. (41)).

Combining the last two results, we conclude

$$\frac{1}{2}|(\mathrm{PUI})\int f dx - \sum_{i\in\mathcal{T}} f(t^i)\int \vartheta_i dx|$$

$$\leq \varepsilon + \frac{1}{2}\varepsilon + \frac{1}{2}|(\mathrm{PUI})\int f dx - \sum_{j\in\mathcal{J}} f(t^j)\int \vartheta_j dx| \leq \frac{5}{2}\varepsilon,$$

which completes the proof.

Remark 3.5. The implication converse to that in Theorem 3.4 also holds, its proof being easy (and analogous to that given in [1], proof of Theorem 2.1 or [2], proof of Theorem 2.4). This implies that the PUI-integral is independent of the choice of the interval I.

Since additivity of the PUI-integral is easily seen, Theorem 3.4 implies additivity of the PU-integral in the usual sense:

$$(\mathrm{PU})\int f_1 dx + (\mathrm{PU})\int f_2 dx = (\mathrm{PU})\int (f_1 + f_2) dx$$

holds provided two of the integrals in the formula exist.

4 Multiplication of PU-integrable functions

Lemma MS enables us to prove a theorem on multiplication of PU-integrable functions.

Theorem 4.1. Let $f : \mathbb{R}^n \to \mathbb{R}$ be a PU-integrable function, G an open bounded set such that supp $f \subset G \subset \mathbb{R}^n$.

Then the integral $(\mathrm{PU})\int \phi f dx$ exists for any function $\phi : G \to \mathbb{R}$ of class $C^{(1)}$ which satisfies

$$||\phi||_1 = \sup\{|\phi(x)| + ||D\phi(x)||;\ x\in G\} < \infty. \tag{42}$$

Moreover, there exists a constant $c_1 = c_1(f) > 0$ such that

$$|(\mathrm{PU})\int \phi f dx| \le c_1||\phi||_1$$

holds for any ϕ satisfying (4.1).

First we will prove the theorem under some additional restrictions on ϕ.

Lemma 4.2. Theorem 4.1 holds if we strengthen (42) to

$$\frac{1}{3} \le \phi(x) \le \frac{2}{3},\ ||D\phi(x)|| \le 1 \text{ for } x \in G. \tag{43}$$

Proof. Given $\varepsilon > 0$, $K \ge K_0$, $L \ge 2L_0$, let us find functions δ, ζ from Definition 2.1 such that

$$|(\mathrm{PU})\int f dx - \sum_{i\in\mathcal{T}} f(t^i)\int \vartheta_i dx| \le \varepsilon \tag{44}$$

for every δ-fine K, L, ζ-regular PU-cover $\theta = \{(t^i, \vartheta_i);\ i \in \mathcal{T}\}$ of supp f. Without loss of generality, let us assume

$$B(t, \delta(t)) \subset G \text{ for } t \in \operatorname{supp} f. \tag{45}$$

Applying Lemma MS with the parameters K, $\frac{1}{2}L$, δ, $\frac{1}{3}\zeta$, $K_1 = 1$, $\eta = \frac{1}{3}$ we find the corresponding gauge δ_1, $\delta_1(x) \le \delta(x)$ for $x \in$ supp f. Again without loss of generality let us further assume

$$|f(t)|.|\phi(x) - \phi(t)| \le \frac{\varepsilon}{m(G)} \text{ for } t \in \operatorname{supp} f,\ x \in B(t, \delta(t)). \tag{46}$$

Let

$$\theta = \{(t^i, \vartheta_i);\ i \in \mathcal{T}\},$$

$$\Lambda = \{(t^j, \vartheta_j);\ j \in \mathcal{J}\}$$

be two PU-covers of supp f which are δ_1-fine and $K, \frac{1}{2}L, \frac{1}{3}\zeta$-regular. By Lemma MS both $\phi\theta$ and $(1-\phi)\Lambda$ are δ_1-fine K, L, ζ-regular systems. Evidently, $\phi\theta \cup (1-\phi)\Lambda$ is a δ_1-fine K, L, ζ-regular PU-cover of supp f, hence

$$|(\mathrm{PU})\int f dx - \sum_{i\in\mathcal{T}} f(t^i)\int \phi\vartheta_i dx - \sum_{j\in\mathcal{J}} f(t^j)\int (1-\phi)\vartheta_j dx| \le \varepsilon.$$

We also have

$$|(\mathrm{PU})\int f dx - \sum_{j\in\mathcal{J}} f(t^j)\int \vartheta_j dx| \le \varepsilon$$

which combined with the previous inequality yields

$$|\sum_{i\in\mathcal{T}} f(t^i)\int \phi\vartheta_i dx - \sum_{j\in\mathcal{J}} f(t^j)\int \phi\vartheta_j dx| \le 2\varepsilon. \tag{47}$$

By (46) we estimate

$$|\sum_{i\in\mathcal{T}} f(t^i)\phi(t^i)\int \vartheta_i dx - \sum_{i\in\mathcal{T}} f(t^i)\int \phi\vartheta_i dx|$$

$$\le \sum_{i\in\mathcal{T}} |f(t^i)| \max\{|\phi(t^i) - \phi(x)|;\ x \in B(t^i,\ \delta(t^i))\} \int \vartheta_i dx \le \varepsilon$$

and similarly for the other term in (47). Consequently,

$$|\sum_{i\in\mathcal{T}} f(t^i)\phi(t^i)\int \vartheta_i dx - \sum_{j\in\mathcal{J}} f(t^j)\phi(t^j)\int \vartheta_j dx \le 4\varepsilon. \tag{48}$$

Since $\varepsilon > 0$ has been arbitrary, existence of (PU) $\int \phi f dx$ follows in the standard manner.

Now fix $\varepsilon = 1$, $K = K_0$, $L = 2L_0$. Let δ_2, ζ_2 correspond to this particular choice according to Definition 2.1.

Let $\delta_3 \le \delta_2$ correspond to K_0, L_0, $\frac{1}{3}\zeta_2$, $K_1 = 1$, $\eta = \frac{1}{3}$ according to Lemma MS. Let $\Lambda = \{(t^j,\ \vartheta_j);\ j \in \mathcal{J}\}$ be a fixed δ_3-fine standard PU-cover of supp f. Then we have by (48)

$$|\sum_{i\in\mathcal{T}} f(t^i)\phi(t^i)\int \vartheta_i dx| \le 4 + |\sum_{j\in\mathcal{J}} f(t^j)\phi(t^j)\int \vartheta_j dx| \tag{49}$$

for any δ_3-fine PU-cover $\theta = \{(t^i,\ \vartheta_i);\ i \in \mathcal{T}\}$ of supp f which is K_0, L_0, ζ_2-regular. Since (PU) $\int f\phi dx$ exists, the sum $\sum_{i\in\mathcal{T}} f(t^i)\phi(t^i)\int \vartheta_i dx$ can be made arbitrarily close to this integral by a proper choice of θ. From (49) we obtain

$$|(\mathrm{PU})\int f\phi dx| \le c_2$$

which proves Lemma 4.2.

To prove Theorem 4.1 it suffices to notice that if ϕ satisfies the assumptions of Theorem 4.1 then

$$\frac{1}{3} + (\phi - \min\phi)/3(\max\phi - \min\phi + \max||D\phi||)$$

satisfies the assumptions of Lemma 4.2 provided ϕ is not identically constant. Since Theorem 4.1 trivially holds for $\phi \equiv \mathrm{const}$, the proof immediately follows.

5 Saks-Henstock Lemma

We will conclude this paper by proving a lemma which plays an important part in the proofs of further results concerning the PU-integral, e.g. differentiability of the integral.

Lemma SH(Saks-Henstock). Let $(\mathrm{PU})\int f dx$ exist. Let $\varepsilon > 0$, $K \geq K_0$, $L \geq 2L_0$ be constants. Find δ, ζ from Definition 2.1 corresponding to the given parameters ε, K, L. Let $\theta = \{(t^i, \vartheta_i);\ i \in \mathcal{I}\}$ be a δ-fine K, L, ζ-regular system such that $t^i \in \operatorname{supp} f$, $i \in \mathcal{T}$.

Then

$$|\sum_{i\in\mathcal{T}}[(\mathrm{PU})\int f\vartheta_i dx - f(t^i)\int \vartheta_i dx]| \leq \varepsilon. \tag{50}$$

Proof. Set $\vartheta(x) = \sum_{i\in\mathcal{T}} \vartheta_i(x)$ and assume there is $\eta > 0$ such that

$$\vartheta(x) \leq 1 - \eta. \tag{51}$$

Let K_1 be such that

$$||D\vartheta(x)|| \leq K_1 \text{ for } x \in \mathbb{R}^n. \tag{52}$$

Let ε, K, L, δ, ζ have the meaning from Lemma SH. Consider Lemma MS with $Q = \operatorname{supp} \vartheta$, η and K_1 from (51) and (52), respectively, and with $\frac{1}{2}L$, $\frac{1}{3}\zeta$ instead of L, ζ. Find the corresponding gauge δ_1.

Choose $\alpha > 0$ arbitrary. Since the function $f(1-\vartheta)$ is PU-integrable by Theorem 4.1, we may assume that δ_1 is so small that

$$|(\mathrm{PU})\int f(1-\vartheta)dx - \sum_{j\in\mathcal{J}} f(t^j)(1-\vartheta(t^j))\int \vartheta_j dx| \leq \alpha \tag{53}$$

holds for every δ_1-fine standard PU-cover of supp f and, moreover,

$$|f(t)|K_1\delta_1(t) \leq \frac{\alpha}{m(I)} \tag{54}$$

where I is a compact interval such that

$$B(t,\ \delta_1(t)) \subset I \text{ for } t \in \operatorname{supp} f.$$

Let $\Lambda = \{(t^j, \vartheta_j);\ j \in \mathcal{J}\}$ be a δ_1-fine standard cover of supp f. Then, by Lemma MS $\theta \cup (1-\vartheta)\Lambda$ is a δ-fine K, L, ζ-regular PU-cover of supp f. Hence

$$|(\mathrm{PU})\int f dx - \sum_{i\in\mathcal{T}} f(t^i)\int \vartheta_i dx - \sum_{j\in\mathcal{J}} f(t^j)\int (1-\vartheta)\vartheta_j dx| \leq \varepsilon. \tag{55}$$

We have

$$\begin{aligned}&\textstyle\sum_{i\in\mathcal{T}}((\mathrm{PU})\int f\vartheta_i dx - f(t^i)\int \vartheta_i dx)\\ &= [\textstyle\sum_{i\in\mathcal{T}}(\mathrm{PU})\int f\vartheta_i dx - (\mathrm{PU})\int f dx + \sum_{j\in\mathcal{J}} f(t^j)(1-\vartheta(t^j))\int \vartheta_j dx]\\ &\quad +[(\mathrm{PU})\textstyle\int f dx - \sum_{i\in\mathcal{T}} f(t^i)\int\vartheta_i dx - \sum_{j\in\mathcal{J}} f(t^j)\int(1-\vartheta)\vartheta_j dx]\\ &\quad +[\textstyle\sum_{j\in\mathcal{J}} f(t^j)\int(1-\vartheta)\vartheta_j dx - \sum_{j\in\mathcal{J}} f(t^j)(1-\vartheta(t^j))\int\vartheta_j dx],\end{aligned}$$

and estimating the first, second and third bracket by (53), (55) and (54) (cf. also (52)), respectively, we conclude

$$|\sum_{i\in\mathcal{T}}((\mathrm{PU})\int f\vartheta_i dx - f(t^i)\int \vartheta_i dx)| \le \alpha + \varepsilon + \alpha.$$

Since $\alpha > 0$ was arbitrarily small, (50) follows immediately.

If (51) is fulfilled for no positive η, we put $\tilde{\vartheta}_i(x) = (1-\eta)\vartheta_i(x)$ obtaining in the same way as above the inequality (50) for $\tilde{\vartheta}_i$ instead of ϑ_i. Dividing it by $1-\eta$ we conclude that (50) holds for ϑ_i with the right hand side $\varepsilon/(1-\eta)$ and, since η can be chosen arbitrarily, (50) holds as well.

References

1 Jarník J., Kurzweil J. : A nonabsolutely convergent integral which admits transformation and can be used for integration on manifolds. Czechoslovak Math. J. 35 (110) (1985), pp. 116-139,

2 Jarník J., Kurzweil J.: A new and more powerful concept of the PU-integral. Czechoslovak Math. J. 38 (113) (1988), pp. 8-48.

Mathematical Institute

Czechoslovak Academy of Sciences

Prague

Czechoslovakia

1-differentials on 1-cells: a further study

Solomon Leader

The Kurzweil-Henstock concept of integral as gauge-limit of approximating sums over tagged-cell divisions should make the traditional approaches to measure and integration on $\mathbb{R}^n$ obsolete. It yields a definition of differential as object of integration that can clarify and expand elementary calculus. The m-differentials on an n-cell introduced and developed in [1] yield rigorous differential formulations for various concepts associated with integration. An expository outline of the case for $m = n = 1$ was presented in [2]. It is this case that we examine further here, although many of our results are valid in higher dimensions.

Let K be a 1-cell $[a, b]$. A ***tagged cell*** (I, t) is a 1-cell I with a selected endpoint t. Our initial objects of integration are "summants" which provide the summands in our approximating sums. A ***summant*** S is a real function $S(I, t)$ on the set of all tagged cells (I, t) in K. Each function x on K defines a summant Δx given by $\Delta x(I, t) = \Delta x(I) = x(q) - x(p)$ for $I = [p, q]$. Traditional summants are of the form $S = z\Delta x$ defined for functions x, z on K by $S(I, t) = z(t)\Delta x(I)$.We shall be interested also in $z|\Delta x|$ and in some novel summants such as $Q(I, t) = 1$ for t a left endpoint, -1 for t a right endpoint of I.

A ***figure*** F is a nonvoid union of finitely many 1-cells in K. A ***division*** $\mathcal{F}$ of F is a finite set of nonoverlapping tagged cells whose union is F. A ***gauge*** α is a function on K with $\alpha(t) > 0$ for all t. (I, t) is ***α-fine*** if the length of I is less than $\alpha(t)$. An ***α-division*** is a division whose members are α-fine. Given a summant S each division $\mathcal{F}$ of F yields the sum $\sum(S, \mathcal{F})$ of $S(I, t)$ over all members (I, t) of $\mathcal{F}$. For each gauge α let $\underline{\sum}(S, \alpha)$ be the infimum, $\overline{\sum}(S, \alpha)$ the supremum, of $\sum(S, \mathcal{F})$ over all α-divisions $\mathcal{F}$ of F.

Define the ***lower integral*** $\underline{\int}_F S = \sup_\alpha \underline{\sum}(S, \alpha)$ and ***upper integral*** $\overline{\int}_F S = \inf_\alpha \overline{\sum}(S, \alpha)$ taken over all gauges α on F. If these two integrals are equal their common value defines $\int_F S$. S is ***integrable*** on F if the integral $\int_F S$ exists and is finite. We shall sometimes use $\int_a^b$ or just $\int$ for $\int_K$.

As a function space the summants on K form a Riesz space (vector lattice) $\mathcal{S}$. The summants with $\int |S| = 0$ form a Riesz ideal $\mathcal{T}$, a linear subspace of $\mathcal{S}$ such that $S \in \mathcal{S}, T \in \mathcal{T}, |S| \leq |T|$ imply $S \in \mathcal{T}$. Thus $\mathcal{D} = \mathcal{S}/\mathcal{T}$ is a Riesz space with the linear and lattice operations transferred homomorphically from $\mathcal{S}$ to $\mathcal{D}$. A ***differential*** σ on K is any element of $\mathcal{D}$. Explicitly σ is an equivalence class $[S]$ of summants under the equivalence $S \sim S'$ defined by $\int |S - S'| = 0$. For $\rho = [R]$ and $\sigma = [S]$ we have $\rho + \sigma = [R + S]$, $c\sigma = [cS]$ for any constant c, and

$|\sigma| = [|S|]$. For the lattice operations these give $\rho \wedge \sigma = [R \wedge S]$, $\rho \vee \sigma = [R \vee S]$, $\sigma^+ = [S^+]$, and $\sigma^- = [S^-]$. The definitions $\underline{\int}\sigma = \underline{\int}S$ and $\overline{\int}\sigma = \overline{\int}S$ are effective for any summants S representing σ. When these two integrals are equal their common value defines $\int \sigma$. σ is ***integrable*** if $\int \sigma$ exists and is finite.

For 1_E the indicator of a subset E of K the definition $1_E\sigma = [1_E S]$ for $\sigma = [S]$ is effective because 1_E is bounded. E is ***σ-null*** if $1_E\sigma = 0$. σ (or any of its representing summants S) is ***tag-null*** if every point is σ-null. A condition holds ***σ-everywhere*** (or at ***σ-all*** points t) in K if it holds on the complement of some σ-null set.

If y is a function defined σ-everywhere on K we can define $y\sigma = [uS]$ where $\sigma = [S]$ and u is any function on K such that $u = y$ σ-everywhere. That this is effective is nontrivial [1].

Each function x on K induces an integrable differential $dx = [\Delta x]$ with $\int_I dx = \Delta x(I)$ for every 1-cell I in K. A differential σ is integrable if and only if $\sigma = dx$ for some x.

A differential σ is ***summable*** if the norm $\nu(\sigma) = \overline{\int}|\sigma|$ is finite. σ is ***tag finite*** if $1_p\sigma$ is summable for every point p. Each representative S of a tag-finite σ effectively defines $S\rho = [SR]$ for every differential $\rho = [R]$. If both ρ and σ are tag-finite then the definition $\sigma\rho = [SR]$ is effective.

The convergence $I \to t$ will always mean that the (I, t) are tagged cells and the length of I goes to 0.

Our work here centers around summability and some conditions related to it. We shall show that the summable differentials on K form a Banach lattice under the norm ν. We shall prove some convergence theorems for products of a differential with Borel functions under appropriate summability conditions. We shall characterize damper-summability, defined below as in [1], and investigate some conditions it implies, namely the archimedean and weakly archimedean properties. We shall give lattice-theoretical formulations for two absolute continuity conditions involving indicator summants. We shall study summant derivatives. Finally we shall give some applications to the calculus: the product formula, iterated integration-by-parts, Taylor's formula, finite expansions of differentials, and a converse to the fundamental theorem of calculus.

1. SUMMABLE DIFFERENTIALS. The summable differentials on K form a Riesz subspace of the Riesz space $\mathcal{D}$ of all differentials on K. Indeed, under the Riesz norm ν they form a Banach lattice. We need only prove completeness.

Proposition 1. Given a sequence of differentials σ_i on K such that $\sum_i \nu(\sigma_i) < \infty$ there exists a summable differential $\sigma = \sum_i \sigma_i$ where the series is convergent in the norm ν.

Proof. We may assume $\sigma_i \neq 0$. Let S_i be a summant representing σ_i.

Choose a decreasing sequence of gauges $\alpha_i > \alpha_{i+1}$ such that

(1) $$\overline{\sum}(|S_i|, \alpha_i) < 2\nu(\sigma_i).$$

Let P_i indicate the α_i-fine tagged cells. That is, $P_i(I,t) = 1$ if (I,t) is α_i-fine, and 0 otherwise. By (1)

(2) $$\sum(|P_i S_i|, \mathcal{K}) < 2\nu(\sigma_i)$$

for every division $\mathcal{K}$ of K. So $|P_i S_i|(I,t) < 2\nu(\sigma_i)$ for every tagged cell (I,t). Thus we can define $S(I,t) = \sum_i P_i S_i(I,t)$ since the defining series is absolutely convergent. Let $\sigma = [S]$, the differential represented by S. For $\mathcal{K}$ any α_j-division of K,

$$\begin{aligned} \textstyle\sum(|S - \sum_{i\le j} S_i|, \mathcal{K}) &= \textstyle\sum(|S - \sum_{i\le j} P_i S_i|, \mathcal{K}) \\ = \textstyle\sum(|\sum_{i>j} P_i S_i|, \mathcal{K}) &\le \textstyle\sum(\sum_{i>j} |P_i S_i|, \mathcal{K} \\ = \textstyle\sum_{i>j} \sum(|P_i S_i|, \mathcal{K}) &< 2\textstyle\sum_{i>j} \nu(\sigma_i) \end{aligned}$$

by (2). So $\nu(\sigma - \sum_{i\le j} \sigma_i) \le 2\sum_{i>j} \nu(\sigma_i) \to 0$ as $j \to \infty$. By the triangle inequality, $\nu(\sigma) \le \sum_i \nu(\sigma_i)$. So σ is summable.

2. LIMITS OF INTEGRABLE DIFFERENTIALS. The next result is elementary but noteworthy because it does not demand absolute integrability. It allows $\nu(\sigma_i) = \nu(\sigma) = \infty$.

Proposition 2. If σ_i is integrable and $\nu(\sigma - \sigma_i) \to 0$ as $i \to \infty$ then σ is integrable and $\int \sigma_i \to \int \sigma$.

Proof. Let $\pi_i = \sigma - \sigma_i$. Since $-\nu(\pi_i) \le \underline{\int}\pi_i \le \overline{\int}\pi_i \le \nu(\pi_i)$ and $\nu(\pi_i) \to 0$, both $\underline{\int}\pi_i$ and $\overline{\int}\pi_i$ converge to 0. Since σ_i is integrable $\underline{\int}\pi_i = \underline{\int}\sigma - \int\sigma_i$ and $\overline{\int}\pi_i = \overline{\int}\sigma - \int\sigma_i$. So $\int\sigma_i$ converges to the finite value $\underline{\int}\sigma = \overline{\int}\sigma$. ○

Proposition 3. If σ_i is absolutely integrable and $\nu(\sigma - \sigma_i) \to 0$ then σ is absolutely integrable and $\int \sigma_i \to \int \sigma$.

Proof. Apply Prop. 2 to σ_i, σ and to $|\sigma_i|$, $|\sigma|$.

3. IMPLICATIONS OF $\overline{\int}\sigma < \infty$ AND $|\overline{\int}\sigma| < \infty$.

Proposition 4. If $\overline{\int}_K\sigma < \infty$ then $\overline{\int}_F\sigma < \infty$ for every figure F in K.

Proof. Let S represent σ. By hypothesis there exists c in $\mathbb{R}$ and a gauge α such that

(3) $$\sum(S, \mathcal{K}) \le c \text{ for every } \alpha\text{-division } \mathcal{K} \text{ of } K.$$

Given complementary figures F, G in K fix an α-division $\mathcal{G}$ of G. For any α-division $\mathcal{F}$ of F the union $\mathcal{K} = \mathcal{F} \cup \mathcal{G}$ is an α-division of K. So by (3)

(4) $$\sum(S, \mathcal{F}) = \sum(S, \mathcal{K}) - \sum(S, \mathcal{G}) \le c - \sum(S, \mathcal{G}).$$

Hence, $\overline{\int}_F \sigma \leq c - \sum(S, \mathcal{G}) < \infty$. ○

Proposition 5. If $|\overline{\int}_K \sigma| < \infty$ then $|\overline{\int}_F \sigma| < \infty$ for every figure F in K.

Proof. For complementary figures F, G Proposition 4 gives $\overline{\int}_F \sigma < \infty$ and $\overline{\int}_G \sigma < \infty$. Thus neither of these integrals can equal $-\infty$ since $\overline{\int}_F \sigma + \overline{\int}_G \sigma = \overline{\int}_K \sigma$ which is finite by hypothesis.

Proposition 6. Let $|\overline{\int}_K \sigma| < \infty$. Given S representing σ and $\varepsilon > 0$ let α be a gauge such that

$$\sum(S, \mathcal{K}) \leq \overline{\int}_K \sigma + \varepsilon \text{ for every } \alpha\text{-division } \mathcal{K} \text{ of } K. \tag{5}$$

Then for each figure F in K

$$\sum(S, \mathcal{F}) \leq \overline{\int}_F \sigma + \varepsilon \tag{6}$$

for every α-division $\mathcal{F}$ of F.

Proof. (5) gives (3) for $c = \overline{\int}_K \sigma + \varepsilon$ which gives (4) in the form $\sum(S, \mathcal{F}) \leq \overline{\int}_K \sigma + \varepsilon - \sum(S, \mathcal{G})$ for all α-divisions $\mathcal{F}, \mathcal{G}$ of complementary figures F, G. That is,

$$\sum(S, \mathcal{F}) \leq \overline{\int}_F \sigma + \varepsilon + [\overline{\int}_G \sigma - \sum(S, \mathcal{G})]$$

with the two upper integrals finite by Proposition 5. The bracketed term can be made arbitrarily small by appropriate choice of $\mathcal{G}$. Hence (6). ○

Proposition 7. Given $\overline{\int}_K \sigma$ finite define $x(t) = \overline{\int}_a^t \sigma$. Then:

(i) for every cell I in K, $\Delta x(I) = \overline{\int}_I \sigma$,

(ii) $dx \geq \sigma$,

(iii) if y is any function on K with $dy \geq \sigma$ then $dy \geq dx$,

(iv) $\underline{\int} dx - \sigma = 0$,

(v) for every bounded function $u \geq 0$ on K, $\underline{\int} u dx \leq \overline{\int} u\sigma \leq \overline{\int} u dx$.

Proof. $x(t)$ is finite by Proposition 5. (i) follows from the additivity of both sides. To get (ii) take any S representing σ. Given $\varepsilon > 0$ take a gauge α satisfying (5). Given an α-division $\mathcal{K}$ of K let $\mathcal{F}$ consist of those members of $\mathcal{K}$ for which $S \geq \Delta x$. Then $\sum((S - \Delta x)^+, \mathcal{K}) = \sum(S - \Delta x, \mathcal{F}) = \sum(S, \mathcal{F}) - \sum(\Delta x, \mathcal{F}) = \sum(S, \mathcal{F}) - \overline{\int}_F \sigma \leq \varepsilon$ by (i) and (6). So $(\sigma - dx)^+ = 0$ which gives (ii). To get (iii) let $dy \geq \sigma$. Then (i) gives $\Delta x(I) \leq \int_I dy = \Delta y(I)$ so $dx \leq dy$. We get (iv) from $\underline{\int} dx - \sigma = \int dx - \overline{\int} \sigma = \Delta x - \overline{\int} \sigma = 0$ by (i). To get the first inequality in (v) let $0 \leq u \leq c$. We may assume $\underline{\int} u dx > -\infty$ and $\overline{\int} u\sigma < \infty$. Then both of these integrals must be finite since $\underline{\int} u dx - \overline{\int} u\sigma \leq \underline{\int} u(dx - \sigma) \leq c \underline{\int} dx - \sigma = 0$ by (iv). This gives the first inequality in (v). The second follows from (ii) for $u \geq 0$.

4. CONVERGENCE THEOREMS FOR SUMMABLE DIFFERENTIALS.

For $dx \geq 0$ a function z on K is *dx-measurable* if given c in $\mathbb{R}$ and

E the set of all t with $z(t) > c$, $1_E dx$ is integrable. (See section 12 on p.173 in [1].) Every Borel function z is dx-measurable.

Proposition 8. Given σ summable on K let $x(t) = \overline{\int}_a^t |\sigma|$ for all t in K. Then $\nu(z\sigma) = \int |z| dx$ for every dx-measurable function z on K.

Proof. Let $z_n = |z| \wedge n$. Then $z_n dx$ is integrable. So $\nu(z_n\sigma) = \int z_n dx$ by (v) in Proposition 7. Hence $\int z_n dx \le \nu(z\sigma) \le \int |z| dx$ since $|\sigma| \le dx$ by Proposition 7. By the monotone convergence theorem (Theorem 10 in [1]) $\int z_n dx \nearrow \int |z| dx$. ○

$\nu(1_E\sigma)$ is the "full variational outer measure" [3] induced by σ on the subsets E of K. Proposition 8 gives $\nu(1_E\sigma) = \int 1_E dx$ for all dx-measurable E, hence for all Borel sets E. So x is the distribution function for the induced Borel measure.

Proposition 9. (Dominated Convergence) Let σ be a differential on K. Let $y_1, y_2 \ldots$ and v be Borel functions on K such that $|y_k| \le v$ for all k in $\mathcal{N}$, $v\sigma$ is summable, and $y_k \to 0$ σ-everywhere. Then $\nu(y_k\sigma) \to 0$ as $k \to \infty$.

Proof. Propositions 7, 8 give x such that $dx \ge v|\sigma|$ and $\nu(zv\sigma) = \int |z| dx$ for every Borel function z on K. Applying this identity with z the indicator of a Borel set we conclude that every σ-null Borel set is dx-null. Similarly the set A of points where $v = 0$ is dx-null since it is $v\sigma$-null. Each $|y_k|/v$ is a Borel function on $K \setminus A$ bounded by 1. Moreover $|y_k|/v \to 0$ except on a σ-null Borel set, hence dx-everywhere. Thus $\nu(y_k\sigma) = \int |y_k/v| dx \to 0$ by Proposition 8 and the bounded convergence theorem (Theorem 10 [1]). ○

Proposition 10. Let σ be a differential on K. Let $v, v_1, v_2, \ldots$ be non-negative Borel functions on K such that $v = \sum_{i \in \mathcal{N}} v_i$ σ-everywhere and $v\sigma$ is summable. Then $\sum_{i \in \mathcal{N}} v_i\sigma$ is ν-convergent to $v\sigma$.

Proof. Apply Proposition 9 with $y_k = v - \sum_{i \le k} v_i$. ○

Note that the required summability of $v\sigma$ is ensured if $\sum_i \nu(v_i\sigma) < \infty$ since this sum dominates $\nu(v\sigma)$ by Theorem 7 [1].

5. DAMPER-SUMMABLE AND ARCHIMEDEAN DIFFERENTIALS. An important consequence of summability is the archimedean property. A member σ of a lattice group is ***archimedean*** if $\pi = 0$ is the only member such that $k|\pi| \le |\sigma|$ for every k in $\mathcal{N}$. Differentials do not have to be summable to be archimedean. Indeed, damper-summable differentials are archimedean. A ***damper*** is a function u with $u(t) > 0$ for all t in K. σ is ***damper-summable*** if $u\sigma$ is summable for some damper u. The next two results relate damper-summability to finiteness of derivates.

Proposition 11. For σ a differential on K the following are equivalent:

(a) There is a representative S of σ and an increasing function x on K such that (i) every dx-null set is σ-null, (ii) at dx-all t $\overline{\lim}|S(I,t)/\Delta x(I)| < \infty$ as $I \to t$.

(b) K is the union of countably many sets E with $1_E\sigma$ summable.

(c) σ is damper-summable.

(d) There is an increasing function x on K such that (i) holds, x is continuous wherever σ is tag-null, and (ii) holds for all $S \in \sigma$.

Proof. (a) $\Rightarrow$ (b). Let E_k consist of all t where $|S| < k\Delta x$ at (I,t) ultimately as $I \to t$. Then $1_{E_k}|\sigma| \leq kdx$. So each $1_{E_k}\sigma$ is summable. By (ii) there is a dx-null set A such that $A, E_1, E_2, \ldots$ cover K. $1_A\sigma$ is summable since it equals 0 by (i).

(b) $\Rightarrow$ (c). We may assume K is covered disjointly by $A_1, A_2, \ldots$ with each $1_{A_k}\sigma$ summable. Choose $a_k > 0$ such that the series $\sum_k a_k\nu(1_{A_k}\sigma) < \infty$. Define the function u on K by letting $u = a_k$ on A_k. $u\sigma$ is summable by Theorem 7 [1].

(c) $\Rightarrow$ (d). Let $w\sigma$ be summable with damper w. Apply Proposition 7 to get $dx \geq w|\sigma|$. Then (i) is trivial since $|\sigma| \leq dx/w$. x is continuous at p if $1_p\sigma = 0$ by Proposition 8 with $z = 1_p$. By adding the identity function to x we may assume that x is everywhere increasing. By adding a unilateral saltus at each point of discontinuity we may assume x is continuous wherever it is either left or right continuous. That is, dx is balanced [1]. Each of these additions preserves (i) and (ii). Take a representative S' of σ with $w|S'| \leq \Delta x$. Given S representing σ let P indicate $\Delta x \leq |S - S'|$. Then $0 \leq P\Delta x \leq |S - S'| \sim 0$. So $Pdx = 0$. By Theorem 16 [1] the indicator summant P is tag-null dx-everywhere. So at dx-all t, $P(I,t) = 0$ ultimately as $I \to t$. Now $P = 0$ means $|S - S'| < \Delta x$ which implies $|S|/\Delta x < 1 + |S'|/\Delta x \leq 1 + 1/w$. Hence (ii) holds for each S representing σ.

(d) $\Rightarrow$ (a) a fortiori. ○

(For higher dimensional cells K the equivalence of (a), (b), (c) remains valid, but the implication (c) $\Rightarrow$ (d) fails.)

Proposition 12. Let y be continuous on K. Then dy is damper-summable if and only if there exists a continuous, increasing function x on K such that (i) every dx-null set is dy-null, and (ii) the derivates of y with respect to x are finite dx-everywhere.

Proof. Apply Proposition 11 to $\sigma = dy$ with $S = \Delta y$.

Our next result does not explicitly demand damper-summability but requires only the archimedian property. A set $\mathcal{P}$ of tagged cells in K is σ-***negligible*** if $P\sigma = 0$ for the indicator P of $\mathcal{P}$.

Proposition 13. Let σ, τ be differentials on K with σ archimedean. Let

z be a function on K. Then $\tau = z\sigma$ if and only if (a) every σ-negligible set of tagged cells is τ-negligible, and (b) given S, T representing σ, τ and $\varepsilon > 0$ the set of tagged cells at which

$$|T - zS| \geq \varepsilon |S| \tag{7}$$

is σ-negligible.

Proof. Let $\tau = z\sigma$. If P is an indicator summant with $P\sigma = 0$ than $P\tau = zP\sigma = 0$. So (a) holds. To get (b) let R indicate (7). Then $\varepsilon R|S| \leq |T - zS|$ so $\varepsilon R|\sigma| \leq |\tau - z\sigma| = 0$. Thus $R\sigma = 0$ which gives (b). Conversely let (a), (b) hold. Choose S, T representing σ, τ. Given $\varepsilon > 0$ let R indicate (7). Then $R\sigma = R\tau = 0$ by (b), (a). Let $Q = 1 - R$. Q indicates $|T - zS| < \varepsilon |S|$. So $Q|\tau - z\sigma| \leq \varepsilon |\sigma|$. Since $Q\sigma = \sigma$ and $Q\tau = \tau$, $Q(\tau - z\sigma) = \tau - z\sigma$. Hence $|\tau - z\sigma| \leq \varepsilon |\sigma|$ for all $\varepsilon > 0$. Thus $\tau = z\sigma$ since σ is archimedean. ○

We remark that (b) is a differentiation condition. It says $S \neq 0$ and for each $\varepsilon > 0$, $|T/S - z| < \varepsilon$ except on a σ-negligible set of tagged cells in K.

6. WEAKLY ARCHIMEDEAN DIFFERENTIALS. Some results on integration of derivatives hold under a condition ostensibly less stringent than archimedean. We call σ ***weakly archimedean*** if $P\sigma = 0$ for every tag-null summant P. Every archimedean differential is weakly archimedean since $|P| \leq \varepsilon$ ultimately under gauge refinement for P tag-null. So $|P\sigma| \leq \varepsilon |\sigma|$ for all $\varepsilon > 0$. Since σ is archimedean this implies $P\sigma = 0$. For summants R, S define $R = o(S)$ to mean *given $\varepsilon > 0$ there exists a gauge α such that $|R| \leq \varepsilon |S|$ at all α-fine tagged cells in K.*

Proposition 14. σ is weakly archimedean if and only if given S representing σ and $R = o(S)$ then $R \sim 0$.

Proof. Let σ be weakly archimedean and $R = o(S)$ for some S representing σ. Define $P = R/S$ if $S \neq 0$, $P = 0$ if $S = 0$. P is tag-null since $R = o(S)$. So $P\sigma = 0$. That is, $R \sim PS \sim 0$. Conversely given P tag-null define $R = PS$. Then $R = o(S)$ so $R \sim 0$, that is, $P\sigma = 0$. ○

We conjecture that there exist weakly archimedean differentials that are not archimedean, and archimedean differentials that are not damper-summable. For weakly archimedean differentials the only conclusion we can draw in the direction of summability is tag-finiteness. σ is ***tag-finite*** if $1_p\sigma$ is summable for every point p in K. [1]

Proposition 15. If σ is weakly archimedean then σ is tag-finite.

Proof. Suppose $\nu(1_p\sigma) = \infty$ for some p. Then given S representing σ there is a sequence $I_k \to p$ such that $|S(I_k, p)| \to \infty$. Let R indicate the set of tagged cells which occur in this sequence. Then $R = o(S)$ so $R \sim 0$ by Proposition 14. But $\overline{\int} R \geq 1$, a contradiction. ○

The next result generalizes Theorem 15 [1] in that "weakly archimedean" replaces "damper-summable", and the demand that x be continuous is dropped.

Proposition 16. Let $x = (x_1, \ldots, x_m)$ be an m-function on K such that each dx_i is weakly archimedean. Let f be a 1-function on a neighborhood of $x(K)$ in $\mathbb{R}^m$ such that for some subset A of K: (i) f is differentiable at $x(t)$ for all t in A, (ii) $1_A dy = dy$ for y defined on K by $y(t) = f(x(t))$. Define the m-function z on K by $z(t) = \text{grad } f(x(t))$ for t in A, $z(t) = 0$ for t in $K \setminus A$. Then dy is weakly archimedean and $dy = z \cdot dx$.

Proof. $||dx||_1 = |dx_1| + \cdots + |dx_m|$ is weakly archimedean. Hence, so is $||dx||$ in the Euclidean norm since all norms on $\mathbb{R}^m$ are equivalent. Clearly $1_A \Delta y - z \cdot \Delta x = o(||\Delta x||)$. So $dy = 1_A dy = z \cdot dx$ by Proposition 14. This implies dy is weakly archimedean since each dx_i is weakly archimedean.

7. SUMMANT DERIVATIVES. We say that the derivative of a summant T with respect to a summant S exists in the *narrow sense* at p with value c if $S(I,p) \neq 0$ ultimately and

$$T(I,p)/S(I,p) \to c \text{ as } I \to p. \tag{8}$$

(Thomson [3] calls this the "ordinary sense".) The derivative exists in the *broad sense* if $S(I,p)$ does not ultimately vanish and (8) holds for $S(I,p) \neq 0$. Our definitions include the cases where $c = \infty$, $c = -\infty$. For c finite the broad sense derivative equals c if and only if $(T - cS)1_p = o(S)$ and $1_p S \neq o(S)$. The latter condition assures uniqueness of c since it gives a (non-void) filterbase for (8). Note that the set E of all p where $1_p S = o(S)$ is σ-null for $\sigma = [S]$ since $1_E S = 0$ ultimately under gauge refinement. The (broad or narrow) derivative of T with respect to S is generally not an invariant of the differentials represented by T, S. With this caution in mind we can retain the traditional notation dy/dx of Leibniz for the (broad or narrow) derivative $\lim \Delta y(I)/\Delta x(I)$ as $I \to t$. We shall similarly use $|dy|/dx, dy/|dx|, |dy|/|dx|$ as in [1]. For weakly archimedean differentials results on derivatives follow readily from Proposition 14. Our first such result generalizes Theorem 18 [1].

Proposition 17. Let σ, τ be differentials on K with σ weakly archimedean. Suppose they have representatives S, T for which the derivative (8) exists in the broad sense and is finite at τ-all points p. Then τ is weakly archimedean. Let A be a subset of K such that $1_A \tau = \tau$ and the finite derivative (8) exists in the broad sense at each p in A. Define $z(p)$ to be this derivative for p in A, and 0 for p in $K \setminus A$. Then $\tau = z\sigma$.

Proof. $1_A T - zS = o(S)$. So $\tau = 1_A\tau = z\sigma$ by Proposition 14. Since σ is weakly archimedean so is $z\sigma$.

Proposition 18. Let σ, τ be differentials on K with σ weakly archimedean. Given representatives S, T let E be the set of all p where the broad sense derivative (8) equals 0. Then E is τ-null.

Proof. $1_E T = o(S)$. Hence $1_E\tau = 0$ by Proposition 14.

Proposition 19. Let σ, τ be differentials on K with σ weakly archimedean. Given representatives S, T then the set E of all p where the broad sense derivative $\lim_{I\to p} |S(I,p)/T(I,p)| = \infty$ is τ-null.

Proof. Given p in E and a positive integer k, $|S| \geq k|T|$ ultimately at (I,p) as $I \to p$. So $1_E T = o(S)$. Thus $1_E\tau = 0$ by Proposition 14. ○

The last three results have interesting consequences for the case where $S = \Delta x$, $T = \Delta y$, $\sigma = dx$, $\tau = dy$. For instance given dx weakly archimedean, Proposition 17 (or 18) implies that if $dy/dx = 0$ in the broad sense dy-everywhere then y is constant. There are conditions under which summant derivatives are differential-invariant almost everywhere justifying the term "differential quotient" for the derivative. A differential σ is ***dampable*** [1] if $u\sigma$ is absolutely integrable for some damper u. σ is ***balanced*** [1] if $1_p\sigma = 0$ wherever there is a tagged cell (I,p) with $\int_I 1_p\sigma$ equal to zero. The next result is valid on 1-cells.

Proposition 20. Let σ be a dampable, balanced differential on the 1-cell K. Let z be a function on K. Given T, S representing $z\sigma, \sigma$ respectively then the narrow sense derivative (8) exists and equals $z(p)$ at σ-all points p in K.

Proof. Given $\varepsilon > 0$ let R indicate $|T - zS| \geq \varepsilon|S|$. $R\sigma = 0$ by Proposition 13. For some damper u, $u|\sigma|$ is integrable and has the same null sets as σ. So Theorem 16 [1] applied to $u|\sigma|$ implies that R is tag-null σ-everywhere. That is, $R(I,t) = 0$ ultimately as $I \to t$ at σ-all t. Now $R = 0$ means $S \neq 0$ and $|T/S - z| < \varepsilon$. Since a countable union of σ-null sets is σ-null the proof is done. ○

For integrable, tag-null σ Thomson [3] proved that $|\sigma|$ dampable implies σ dampable, answering a question posed in [2]. An alternate proof comes from the following differentiation theorem.

Proposition 21. Let x be a continuous function on the 1-cell K with $|dx|$ dampable. Then the narrow sense derivative $dx/|dx|$ exists dx-everywhere (with values 1 or -1 of course).

Proof. (All derivatives here are narrow sense.) Given a damper z for $|dx|$ there is an isotone, continuous function y on K such that $dy = z|dx|$. By Proposition 20 with $\sigma = |dx|$, $T = \Delta y$, and $S = |\Delta x|$

$$dy/|dx| = z > 0 \tag{9}$$

dx-everywhere. There exists a gauge α on K such that at each t where (9) holds both $\Delta y(I) > 0$ and $|\Delta x(I)| > 0$ for all α-fine (I, t). Since x is continuous the intermediate value property implies that as $I \to t-$ either $\Delta x(I) > 0$ ultimately or $\Delta x(I) < 0$ ultimately. A similar statement holds as $I \to t+$. So (9) implies the existence of the left and right derivatives $(dx/dy)_-, (dx/dy)_+$ with values restricted to $\pm 1/z$. These unilateral derivatives can differ only at countably many points ([4]p.359). A countable set is dx-null since dx is tag-null by continuity of x. So dx/dy exists and is finite dx-everywhere. Since (9) also holds dx-everywhere so does $dx/|dx| = z dx/dy$. ○

We can now get Thomson's result.

Proposition 22. Let x be continuous on the 1-cell K with $|dx|$ dampable. Then dx is dampable and every damper for $|dx|$ is a damper for dx.

Proof. Extend x to the right of $K = [a, b]$ by defining $x(s) = x(b)$ for $s > b$. Define $w(t) = \overline{\lim}_{n\to\infty} \operatorname{sgn}[x(t + 1/n) - x(t)]$ for $a \leq t \leq b$. Since x is continuous w is a Borel function. Also $|w| \leq 1$. Let z be a damper for $|dx|$. By Proposition 21 $dx/|dx| = w$ dx-everywhere. So $dx = w|dx|$ by Proposition 17. Hence $zdx = wz|dx| = wdy$. Since $dy \geq 0$ and w is a bounded Borel function wdy is integrable. That is, zdx is integrable. ○

8. ABSOLUTE CONTINUITY CONDITIONS INVOLVING INDICATOR SUMMANTS.

For $\sigma, \tau \geq 0$ in any lattice group $\mathcal{D}$ the condition

$$\wedge_{k\in\mathcal{N}}(\tau - k\sigma)^+ = 0$$

implies the condition

(10) $\qquad \rho \wedge \tau = 0$ for all ρ in $\mathcal{D}$ such that $\rho \wedge \sigma = 0$.

The converse holds if σ is archimedean. For the Riesz space $\mathcal{D}$ of differentials on K (10) is just the absolute continuity (a) in Proposition 13. We prove this next.

Proposition 23. For differentials $\sigma, \tau \geq 0$ on K (10) is equivalent to

(11) $\qquad$ Every σ-negligible set of tagged cells is τ-negligible.

Proof. Given (11) and $\rho \wedge \sigma = 0$ pick representatives R, S of ρ, σ with $R \wedge S = 0$. This can be done by replacing arbitrary representatives R, S by $(R-S)^+$, $(R-S)^-$ since $\rho = \rho - \rho\wedge\sigma = (\rho-\sigma)^+$ and $\sigma = \sigma = \sigma - \rho\wedge\sigma = (\rho-\sigma)^-$. Let P indicate $R > 0$. Then $PS = 0$ so $P\sigma = 0$. By (11) $P\tau = 0$. So we can choose $T \geq 0$ representing τ with $PT = 0$. Since $PR = R$, $R \wedge T = (PR) \wedge T = R \wedge (PT) = 0$. So $\rho \wedge \tau = 0$. Hence , (11) implies (10). Conversely, given (10) and an indicator summant P with $P\sigma = 0$ we contend $P\tau = 0$. Since $(P\tau) \wedge \sigma = \tau \wedge (P\sigma) = \tau \wedge 0 = 0$, $P\tau = (P\tau) \wedge \tau = 0$ by (10). ○

For $\sigma, \tau \geq 0$ the ν-convergence of $\tau \wedge (k\sigma)$ to τ as $k \to \infty$ is stronger than (11). Since $\tau - \tau \wedge (k\sigma) = (\tau - k\sigma)^+$ it is just

(12) $$\nu(\tau - k\sigma)^+ \to 0 \text{ as } k \to \infty.$$

Proposition 24. Let $\sigma, \tau \geq 0$ be differentials on K. Then (12) holds if and only if both of the following conditions hold:

(13) $$(\tau - m\sigma)^+ \text{ is summable for some } m \text{ in } \mathcal{N},$$

(14) $$\text{given } \varepsilon > 0 \text{ there exists } \delta > 0 \text{ such that}$$

$$\nu(P\tau) < \varepsilon \text{ for every indicator summant } P \text{ with } \nu(P\sigma) < \delta.$$

Proof. Let (12) hold. Then (13) is trivial. To get (14) apply $0 \leq \tau \leq (\tau - k\sigma)^+ + k\sigma$ to get $0 \leq P\tau \leq (\tau - k\sigma)^+ + kP\sigma$. Hence

(15) $$\nu(P\tau) \leq \nu(\tau - k\sigma)^+ + k\nu(P\sigma).$$

Given $\varepsilon > 0$ apply (12) to get k with $\nu(\tau - k\sigma)^+ < \varepsilon/2$. Take $\delta = \varepsilon/2k$. Then (14) follows from (15). Conversely, let (13) and (14) hold. Choose m from (13). Let $S, T \geq 0$ represent σ, τ. Let P_k indicate $T > kS$. Then

(16) $$0 \leq (T - kS)^+ = P_k(T - kS) = P_kT - kP_kS.$$

So $kP_kS \leq P_kT$, hence $0 \leq kP_k\sigma \leq P_k\tau$. Thus

$$(k - m)P_k\sigma \leq P_k\tau - mP_k\sigma = P_k(\tau - m\sigma) \leq (\tau - m\sigma)^+.$$

So

$$0 \leq (k - m)\nu(P_k\sigma) \leq \nu(\tau - m\sigma)^+ < \infty$$

for $k \geq m$. Thus $\nu(P_k\sigma) \to 0$ as $k \to \infty$. So $\nu(P_k\tau) \to 0$ by (14). Now $(\tau - k\sigma)^+ \leq P_k\tau$ by (16). So $\nu(\tau - k\sigma)^+ \leq \nu(P_k\tau)$ giving (12). ○

9. THE DIFFERENTIAL OF A PRODUCT, ITERATED INTEGRATION-BY-PARTS, AND THE TAYLOR FORMULA. For each tagged cell (I, t) define the summant $Q(I, t)$ to be 1 if t is the left endpoint of I, -1 if t is the right endpoint. If $t + h$ is the endpoint of I opposed to t then $Q(I, t) = \operatorname{sgn} h$. So

(17) $$\Delta x(I) = Q(I, t)[x(t + h) - x(t)] \text{ for } x \text{ a function on } I.$$

Proposition 25. For u, v functions on the 1-cell K

$$\Delta(uv) = u\Delta v + v\Delta u + Q\Delta u\Delta v. \tag{18}$$

If u, v are bounded then

$$d(uv) = udv + vdu + Qdudv. \tag{19}$$

If u is continuous and dv weakly archimedean then

$$d(uv) = udv + vdu. \tag{20}$$

Proof. In the identity

$$u(s)v(s) - u(t)v(t) =$$

$$u(t)[v(s) - v(t)] + v(t)[u(s) - u(t)] + [u(s) - u(t)][v(s) - v(t)]$$

set $s = t + h$, multiply through by $Q = \operatorname{sgn} h$, and apply (17) to get (18). Since $dudv$ is well defined for u and v bounded, (18) gives (19). For u continuous du is tag-null. So $\Delta u\Delta v \sim 0$ for dv weakly archimedean giving (20) from (18). ○

Validity of (20) and integrability of udv imply integrability of vdu through integration-by-parts. For example, if u is continuous and v is of bounded variation then vdu is integrable. In particular, if u is continuous then vdu is integrable for every polynomial function v. Equivalently, every continuous u has finite moments $\int_a^b t^k du(t)$ for $k = 0, 1, \ldots$.

For u, v sufficiently smoothly differentiable (20) extends to the iterated integration-by-parts formula (22).

Proposition 26. Let u, v be functions on the 1-cell K both of class C^n for some $n \geq 1$. Define the continuous function

$$w = \sum_{j=0}^{n} (-1)^j u^{(j)} v^{(n-j)}. \tag{21}$$

Then

$$dw = udv^{(n)} + (-1)^n vdu^{(n)}. \tag{22}$$

Proof. For $0 \leq j < n$ Proposition 17 gives $du^{(j)} = u^{(j+1)}dt$. Thus, since $u^{(j+1)}$ is continuous, $du^{(j)}$ is absolutely integrable, hence archimedean. Also, $v^{(n-j)}$ is continuous. Similarly $dv = v^{(1)}dt$ is archimedean and $u^{(n)}$ is continuous. So by Proposition 25 the classical product rule (20) applies to each term on the right side of (21). Thus

$$\begin{aligned} dw &= \textstyle\sum_{j=0}^{n} (-1)^j d(u^{(j)} v^{(n-j)}) \\ &= \textstyle\sum_{j=0}^{n} (-1)^j u^{(j)} dv^{(n-j)} + \sum_{j=0}^{n} (-1)^j v^{(n-j)} du^{(j)} \\ &= udv^{(n)} + [\textstyle\sum_{j=1}^{n} (-1)^j u^{(j)} v^{(n-j+1)} + \\ & \textstyle\sum_{j=0}^{n-1} (-1)^j u^{(j+1)} v^{(n-j)}]dt + (-1)^n vdu^{(n)} \end{aligned}$$

The quantity in brackets vanishes since for $1 \leq j \leq n$ term j in the first sum negates term $j-1$ in the second sum. Hence (22). ○

Proposition 26 gives a general Taylor formula.

Proposition 27. Let u be a function of class C^n on a 1-cell J containing the points p, q. Then

$$u(q) = \sum_{j=0}^{n} \frac{u^j(p)}{j!}(q-p)^j + r_n \tag{23}$$

where

$$r_n = \int_p^q \frac{(q-t)^n}{n!} du^{(n)}(t). \tag{24}$$

Thus

$$|r_n| \leq \frac{|q-p|^n}{n!} \int_J |du^{(n)}|. \tag{25}$$

Proof. Define $v(t) = (t-q)^n/n!$. By repeated differentiation

$$v^{(n-j)}(t) = (t-q)^j/j! \text{ for } j = 0, \ldots, n. \tag{26}$$

In particular $v^{(n)}(t) = 1$, so $dv^{(n)} = 0$. Hence (22) in Prop 26 reduces to $dw = (-1)^n v du^{(n)}$. Integrating this from p to q we get $w(q) - w(p) = r_n$ with the implied existence of the integral $\int_p^q (-1)^n v(t) du^{(n)}(t)$ in (24) defining r_n. So $w(q) = w(p) + r_n$. This is just (23). Indeed $w(q) = u(q)$ by (21) since $v^{(n-j)}(q) = 0$ for $0 < j \leq n$ by (26). $w(p)$ is just the sum in (23) by (21),(26). (25), of use for $u^{(n)}$ of bounded variation, comes from (24). ○

10. DIFFERENTIAL FORMULAS FOR FUNCTIONALLY RELATED VARIABLES. For x bounded (19) gives $d(x^2) = 2x dx + Q(dx)^2$. This is a special case, $u(x) = x^2$, of our next result. For brevity we introduce the notation $\tilde{\Delta}x(I,t) = Q(I,t)\Delta x(I)$ and $\tilde{d}x = Qdx$. Since $Q^2 = 1$ we have $\Delta x = Q\tilde{\Delta}x$ and $dx = Q\tilde{d}x$. Also $\tilde{d}x = [\tilde{\Delta}x]$.

Proposition 28. Let x be a bounded function on K and $y = u(x)$ where u is a polynomial of degree $n > 0$. Then

$$\tilde{d}y = \sum_{j=1}^{n} \frac{u^{(j)}(x)}{j!} (\tilde{d}x)^j \tag{27}$$

Proof. The Taylor identity

$$u(x+h) - u(x) = \sum_{j=1}^{n} \frac{u^{(j)}(x)}{j!} h^j$$

with $x = x(t)$ and $h = \tilde{\Delta}x(I,t)$ gives

$$\tilde{\Delta}y(I,t) = \sum_{j=1}^{n} \frac{u^{(j)}(x)}{j!}(\tilde{\Delta}x(I,t))^j.$$

This gives (27) since x is bounded, making $(\tilde{dx})^j$ well defined. ○

With suitable restrictions on x we can get (27) for functions u more general than polynomials. Our next result does this.

Proposition 29. Let x be a continuous function on K such that $(dx)^n$ is weakly archimedean for some $n \geq 1$. Let $y = u(x)$ where u is a function on $x(K)$ of class C^n with $u^{(n)}$ of bounded variation. Then (27) holds.

Proof. Given I with endpoints $t, t+h$ let $p = x(t)$, $q = x(t+h)$. Then $\tilde{\Delta}x(I,t) = q - p$ and $\tilde{\Delta}y(I,t) = u(q) - u(p)$. So Proposition 27 gives

$$|\tilde{\Delta}y(I,t) - \sum_{j=1}^{n} \frac{u^{(j)}(p)}{j!}(\tilde{\Delta}x(I,t))^j| \leq |\Delta x(I)|^n \int_J |du^{(n)}| \tag{28}$$

where J is the interval joining q, p. As $I \to t$, $h \to 0$ and $q \to p$ by continuity of x. That is, $J \to p$. Therefore $\int_J |du^{(n)}| \to 0$ since the variation of a continuous function is continuous. So (28) gives

$$\tilde{\Delta}y - \sum_{j=1}^{n} \frac{u^{(j)}(x)}{j!}(\tilde{\Delta}x)^j = o(|\Delta x(I)|^n).$$

This gives (27) by Proposition 14 since $(dx)^n$ is weakly archimedean. ○

11. GETTING $dy/dx = z$ FROM $dy = zdx$. Our final proposition generalizes the classical result that

$$\frac{d}{dx}\int_a^x z(t)dt = z(x)$$

for z continuous. It is strong enough to yield L'Hôpital's rule. The condition (i) is trivial for x strictly monotone on each side of c.

Proposition 30. Let x, y, z be functions on K and c a point in K such that: (i) x is of bounded variation and its variation function v has finite derivates in the narrow sense with respect to x at c, (ii) z is continuous at c, and (iii) $dy = zdx$. Then dy/dx exists in the narrow sense at c and equals $z(c)$.

Proof. By (i) there exists $M < \infty$ such that

$$\Delta v(I) < M|\Delta x(I)| \tag{29}$$

ultimately as $I \to c$ in K. Let $p = z(c)$. Given $\varepsilon > 0$ choose $\delta > 0$ such that for all I with (I, c) δ-fine (29) holds and by (ii)

$$|z - p| < \varepsilon/M \text{ on } I. \tag{30}$$

For such I,

$$\begin{aligned} |(\Delta y - p\Delta x)(I)| &= |\textstyle\int_I dy - p dx| &&= |\textstyle\int_I (z-p) dx| \\ &\leq (\varepsilon/M) \textstyle\int_I dv &&= (\varepsilon/M)\Delta v(I) \\ &< \varepsilon|\Delta x(I)| \end{aligned}$$

by(30) and (29). That is, $|\Delta y(I)/\Delta x(I) - p| < \varepsilon$. ○

REFERENCES

1. S. Leader, A concept of differential based on variational equivalence under generalized Riemann integration, Real Analysis Exchange 12 (1986-87), 144-175.

2. ———, What is a differential? A new answer from the generalized Riemann integral, Amer. Math. Monthly 93 (1986), 348-356.

3. B.S. Thomson, Some remarks on differential equivalence, Real Analysis Exchange 12 (1986-87), 294-312.

4. E.C. Titchmarsh, The Theory of Functions, London (1950).

Rutgers University
New Brunswick
New Jersey 08903
U.S.A.

Generalized convergence theorems for Denjoy-Perron integrals

Lee Peng Yee

1. INTRODUCTION

Djvarsheishvili [2 page 50, 4] proved a generalized convergence theorem for the Denjoy integral. Lee and Chew [9,10,11] also proved independently a series of generalized convergence theorems for the Henstock integral, one of which is equivalent to Djvarsheishvili's result. It is well-known that the Denjoy and Henstock integrals are equivalent [5,8,13]. At present, there are essentially three different proofs of the generalized convergence theorems. One is by a kind of linearization (Theorem 2 below), another by the standard category argument which is commonly used in classical integration theory, and a third proof by means of the Henstock integral. The first two proofs rely heavily on the real-line properties. A part of the third proof also depends on a real-line property.

To free the theory from its one-dimensional setting, we have to provide a proof that is real-line independent. We shall present such a proof in this note. For simplicity, we present it in detail on the real line and indicate how to extend it to the n-dimensional space. Some applications will also be mentioned in Section 4.

2. MAIN THEOREMS

We shall state the main theorem here and prove it in the next section. A general form will apear in Section 5. First, we introduce some necessary terms.

Let $[a,b]$ be a compact interval. A division D given by

$$a = x_0 < x_1 < \ldots < x_n = b \text{ and } \{\xi_1, \xi_2, \ldots, \xi_n\}$$

is said to be δ-fine if there is a $\delta(\xi) > 0$ such that $x_{i-1} \le \xi_i \le x_i$ and $[x_{i-1}, x_i] \subset (\xi_i - \delta(\xi_i), \xi_i + \delta(\xi_i))$ for all i. We call $\{x_i\}$ the division points and ξ_i the associated point of $[x_{i-1}, x_i]$. For brevity, we write $D = \{[u,v];\xi\}$ where $[u,v]$ is a typical interval in D and ξ the associated point of $[u,v]$. If D is δ-fine, then $\xi \in [u,v] \subset (\xi - \delta(\xi)),\ \xi + \delta(\xi))$. Next, let $X \subset [a,b]$ and very often X is closed. A division D of X is defined to be a family, denoted again by $\{[u,v];\xi\}$, of a finite number of non-overlapping intervals $[u,v]$ with $\xi \in [u,v]$ and $\xi \in X$

such that the union of $[u,v]$ covers X. A subset of the division of X is called a partial division of X.

Let $D_1 = \{[u,v];\xi\}$ and $D_2 = \{[s,t];\eta\}$ be two partial divisions of X. Let E_1 denote the union of intervals $[u,v]$ in D_1 and E_2 the union of intervals $[s,t]$ in D_2. Then $D_1 \setminus D_2$ denotes the collection of component intervals in $E_1 - E_2$. A real-valued function F is said to be $AC^{**}(X)$ with X being a closed set in $[a,b]$ if for every $\varepsilon > 0$ there exist a $\delta(\xi) > 0$ and an $\eta > 0$ such that for any two δ-fine partial divisions of X, D_1 and D_2, in which D_2 may be void, we have

$$(D_1 \setminus D_2)\sum |v-u| < \eta \text{ implying } (D_1 \setminus D_2)\sum |F(v)-F(u)| < \varepsilon.$$

A function F is said to be ACG^{**} if $[a,b]$ is the union of a sequence of closed sets X_i, $i = 1,2,\ldots$, on each of which F is $AC^{**}(X_i)$. A sequence of functions $\{F_n\}$ is said to be $UAC^{**}(X)$ if $\delta(\xi) > 0$ and $\eta > 0$ in the definition of $AC^{**}(X)$ are independent of n. Furthermore, $\{F_n\}$ is $UACG^{**}$ if $[a,b]$ is the union of closed sets X_i such that $\{F_n\}$ is $UAC^{**}(X_i)$ for each i.

To understand what $AC^{**}(X)$ means, we refer to the classical definition. A function F is said to be $AC^{*}(X)$ if for every $\varepsilon > 0$ there is an $\eta > 0$ such that for any sequence of non-overlapping intervals $\{[a_k,b_k]\}$ with $a_k, b_k \in X$ we have

$$\sum_k |b_k - a_k| < \eta \text{ implying } \sum_k \omega(F;[a_k,b_k]) < \varepsilon$$

where ω denotes the oscillation of F over $[a_k,b_k]$. Chew pointed out that if F is continuous and X closed then the above definition is equivalent to the one with one point a_k or $b_k \in X$ and with the oscillation replaced by the difference $|F(b_k)-F(a_k)|$. Indeed, this is what Henstock defines in [7] using one endpoint only, and he restricts further to δ-fine intervals for some $\delta(\xi) > 0$. Similarly, we define ACG^{*}. For the purpose of describing the primitive fnnctions, ACG^{*} and ACG^{**} are equivalent as we shall see later. In general , we cannot deduce from the one-point definition that the infinite sum of the oscillations of F over the contiguous intervals of X is finite, assuming X is closed. Hence we introduce $D_1 \setminus D_2$ above in which D_2 may be void. This is to take care of the oscillation of F outside X.

For completeness, we define the Henstock integral as follows [7]. A function f is said to be Henstock integrable to A on $[a,b]$ if for every $\varepsilon > 0$ there is a $\delta(\xi) > 0$ such that for any δ-fine division $D = \{[u,v];\xi\}$ we have

$$|(D)\sum f(\xi)(v-u) - A| < \varepsilon.$$

The prefix (D) is sometimes omitted if there is no ambiguity. If F denotes the primitive of f, we often write $F(u,v) = F(v) - F(u)$. Obviously, $A = F(a,b)$. If both the function f and its absolute value $|f|$ are Henstock integrable then we say that f is absolutely Henstock integrable.

Now we state the controlled convergence theorem.

Theorem 1. Let the following conditions be satisfied:

(i) $f_n(x) \to f(x)$ almost everywhere in $[a,b]$ where each f_n is Henstock integrable on $[a,b]$;

(ii) the primitives F_n of f_n are $UACG^{**}$.

Then f is Henstock integrable on $[a,b]$ and

$$\int_a^b f_n \to \int_a^b f.$$

We remark that in view of $UACG^{**}$ the sequence $\{F_n\}$ is equicontinuous. Then $\{F_n\}$ is uniformly bounded and by Ascoli's theorem it contains a uniformly convergent subsequence $\{F_{n(i)}\}$. Therefore

$$F(x) = \lim_{i\to\infty} F_{n(i)}(x) \text{ exists for } x \in [a,b]$$

and it is continuous on $[a,b]$. It remains to show that F is the primitive of f on $[a,b]$.

3. PROOF

We shall prove Theorem 1 in a series of lemmas.

Lemma 1. Let f_n, $n = 1,2,\ldots$, be Henstock integrable on $[a,b]$ with the primitives F_n, $n = 1,2,\ldots$, $f_n(x) \to f(x)$ almost everywhere in $[a,b]$ as $n \to \infty$, and $F_n(x)$ converges pointwise to a limit function $F(x)$. Then in order that f is Henstock integrable on $[a,b]$ with the primitive F, it is necessary and sufficient that for every $\varepsilon > 0$ there exists $M(\xi)$ taking integer values such that for infinitely many $m(\xi) \geq M(\xi)$ there is a $\delta(\xi) > 0$ such that for any δ-fine division $D = \{[u,v];\xi\}$ we have

$$|\sum F_{m(\xi)}(u,v) - F(a,b)| < \varepsilon.$$

Proof. For simplicity, we may assume that $f_n(x) \to f(x)$ everywhere as $n \to \infty$. Supose f is Henstock integrable on $[a,b]$ with the primitive F. Given $\varepsilon > 0$ and $\xi \in [a,b]$, there is an integer $M(\xi)$ such that whenever $m(\xi) \geq M(\xi)$ we have

$$|f_{m(\xi)}(\xi) - f(\xi)| < \varepsilon.$$

Since each f_n is also Henstock integrable on $[a,b]$, there is a $\delta_n(\xi) > 0$ such that for any δ_n-fine division $D = \{[u,v];\xi\}$ we have

$$\sum |F_n(u,v) - f_n(\xi)(v-u)| < \varepsilon 2^{-n}.$$

Also, there is a $\delta_0(\xi) > 0$ such that for any δ_0-fine division $D = \{[u,v];\xi\}$ we have

$$\sum |F(u,v) - f(\xi)(v-u)| < \varepsilon.$$

Now for every $m(\xi) \geq M(\xi)$, put $\delta(\xi) = \min\{\delta_{m(\xi)}(\xi), \delta_0(\xi)\}$. It follows that for any δ-fine division $D = \{[u,v];\xi\}$ we have

$$\begin{aligned} |\sum F_{m(\xi)}(u,v) - F(a,b)| &\leq \sum |F_{m(\xi)}(u,v) - f_{m(\xi)}(\xi)(v-u)| \\ &\quad + \sum |f_{m(\xi)}(\xi) - f(\xi)|(v-u) \\ &\quad + \sum |f(\xi)(v-u) - F(u,v)| \\ &< \varepsilon + \varepsilon(b-a) + \varepsilon. \end{aligned}$$

That is, the required condition holds.

Conversely, suppose the condition is satisfied. Using the same notation, we choose $m(\xi) \geq M(\xi)$ such that

$$|f_{m(\xi)}(\xi) - f(\xi)| < \varepsilon.$$

Then modify $\delta(\xi)$ so that $\delta(\xi) \leq \delta_{m(\xi)}(\xi)$ for $\xi \in [a,b]$. Then for any δ-fine division $D = \{[u,v];\xi\}$ we have

$$\begin{aligned} |\sum f(\xi)(v-u) - F(a,b) &\leq \sum |f(\xi) - f_{m(\xi)}(\xi)|(v-u) \\ &\quad + \sum |f_{m(\xi)}(\xi)(v-u) - F_{m(\xi)}(u,v)| \\ &\quad + |\sum F_{m(\xi)}(u,v) - F(a,b)| \\ &< \varepsilon(b-a+2). \end{aligned}$$

Hence f is Henstock integrable to $F(a,b)$ on $[a,b]$.

We remark that it is not enough for the condition in Lemma 1 to hold for a single $M(\xi)$ only. It must hold for infinitely many $m(\xi) \geq M(\xi)$ so that when proving the sufficiency we may choose a suitable $f_{m(\xi)}(\xi)$ which is close to $f(\xi)$.

Lemma 2. Let f be Henstock integrable on $[a,b]$ with the primitive F. If F is $AC^{**}(X)$ with X closed in $[a,b]$ then f_X is absolutely Henstock integrable on $[a,b]$ where $f_X(x) = f(x)$ when $x \in X$ and 0 otherwise.

Proof. We shall prove that $f^* = \max(f_X, 0)$ is Henstock integrable on $[a,b]$. If so, then both $f_X = \max(f_X,0) - \max(-f_X,0)$ and $|f_X| = \max(f_X,0) + \max(-f_X,0)$ are Henstock integrable on $[a,b]$.

Let ξ be the associated point of $[u,v]$ and put $F^*(u,v) = \max\{F(u,v),0\}$ when $\xi \in X$ and 0 otherwise. Since f is Henstock integrable on $[a,b]$, for every $\varepsilon > 0$ there is a $\delta(\xi) > 0$ such that for any δ-fine division $D = \{[u,v];\xi\}$ of $[a,b]$ we have

$$\sum |f(\xi)(v-u) - F(u,v)| < \varepsilon.$$

Next define

$$\chi(x,y) = \sup \sum |f(\xi)(v-u) - F(u,v)|$$

where the supremum is taken over all δ-fine divisions $D = \{[u,v];\xi\}$ of $[x,y]$. Then for any δ-fine division $D = \{[u,v];\xi\}$ of $[a,b]$ we have

$$\begin{aligned} f(\xi)(v-u) &\leq F(u,v) + \chi(u,v) \\ &\leq F^*(u,v) + \chi(u,v) \end{aligned}$$

which gives

$$f^*(\xi)(v-u) \leq F^*(u,v) + \chi(u,v).$$

Similarly, we can show that

$$F^*(u,v) \leq f^*(\xi)(v,u) + \chi(u,v).$$

Consequently, we obtain

$$|\sum\{f^*(\xi)(v-u) - F^*(u,v)\}| \leq \sum \chi(u,v) < \varepsilon.$$

Since F is $AC^{**}(X)$, we can define

$$A_\delta = \sup(D)\sum F^*(u,v) \text{ and } A = \inf_\delta A_\delta$$

where the supremum is over all δ-fine divisions D of X. Note that A_δ is finite for $\delta(\xi) > 0$ as given in the definition of $AC^{**}(X)$. Indeed, X is covered by a finite number, say N, of intervals each of length less than η where η comes from the definition $AC^{**}(X)$ with D_2 being void. Then $A_\delta \leq N\varepsilon$. We may assume the $\delta(\xi)$ here is the same as $\delta(\xi)$ above. Choose an open set $G \supset X$ such that $|G - X| < \eta$ where η again comes from the definition of $AC^{**}(X)$. Modify $\delta(\xi)$ if necessary so that $(\xi - \delta(\xi), \xi + \delta(\xi)) \subset G$ when $\xi \in X$ and $(\xi - \delta(\xi), \xi + \delta(\xi))$ does not intersect X when otherwise. Fix a δ-fine division D_1 of X such that

$$A_\delta - \varepsilon < (D_1)\sum F^*(u,v) \leq A_\delta.$$

Let D be another δ-fine division of X which is finer than D_1. IF E_1 is the union of intervals in D_1 and E the union of intervals in D, then $E_1 - E \subset G - X$. This is so because $E_1 \subset G$ and E is a finite cover of X, therefore $E_1 - E$ does not intersect X, otherwise E will not cover X. It follows that

$$(D_1 \setminus D)\sum |v-u| < \eta$$

which implies

$$(D_1 \setminus D)\sum |F(u,v)| < \varepsilon.$$

Then we have

$$\begin{aligned} |(D)\sum f^*(\xi)(v-u) - A_\delta| &\leq |(D)\sum\{f^*(\xi)(v-u) - F^*(u,v)\}| \\ &\quad + A_\delta - (D_1)\sum F^*(u,v) \\ &\quad + (D_1 \setminus D)\sum |F(u,v)| \\ &< 3\varepsilon. \end{aligned}$$

Further, given $\varepsilon > 0$ there is a $\delta(\xi) > 0$ such that

$$A \leq A_\delta < A + \varepsilon.$$

Again, we may assume this $\delta(\xi)$ to be the same as $\delta(\xi)$ above. Finally, define $\delta_1(\xi) \le \delta(\xi)$ such that any δ_1-fine division D of X is finer than D_1. Then for any δ_1-fine division $D = \{[u,v];\xi\}$ of $[a,b]$ we have

$$|\sum f^*(\xi)(v-u) - A| \le |\sum f^*(\xi)(v-u) - A_\delta| + A_\delta - A < 4\varepsilon.$$

Hence f^* is Henstock integrable on $[a,b]$.

Lemma 3. Let the following conditions be satisfied:

(i) $f_{n,X}(X) \to f_X(x)$ almost everywhere in $[a,b]$ as $n \to \infty$ where each $f_{n,X}$ is Henstock integrable on $[a,b]$;

(ii) the primitives $F_{n,X}$ of $f_{n,X}$ are $UAC^{**}(X)$.

Then f_X is Henstock integrable on $[a,b]$ with the primitive F_X and

$$F_{n,X}(a,b) \to F_X(a,b).$$

Proof. In view of Lemma 1, it is sufficient to prove that

$$F_X(x) = \lim_{n\to\infty} F_{n,X}(x) \text{ exists for } x \in [a,b]$$

and that the condition in Lemma 1 holds with $F_{m(\xi)}$ and F replaced respectively by $F_{m(\xi),X}$ and F_X. By (ii), for every $\varepsilon > 0$ there exist a $\delta(\xi) > 0$ and an $\eta > 0$, both independent of n, such that for any δ-fine partial division D of X satisfying

$$(D)\sum |v-u| < \eta \text{ we have } (D)\sum |F_{n,X}(u,v)| < \varepsilon.$$

By Egoroff's theorem, there is an open set G with $|G| < \eta$ such that

$$|f_n(\xi) - f_m(\xi)| < \varepsilon \text{ for } n,m \ge N \text{ and } \xi \notin G.$$

Consider the following, in which D is a δ-fine division of $[x,y]$ and $D = D_1 \cup D_2$ so that D_1 contains the intervals with the associated points $\xi \notin G$ and D_2 otherwise,

$$\begin{aligned} |F_{n,X}(x,y) - F_{m,X}(x,y)| &= |(D)\sum\{F_{n,X}(u,v) - F_{m,X}(u,v)\}| \\ &\le (D_1)\sum |F_{n,X}(u,v) - f_{n,X}(\xi)(v-u)| \\ &\quad + (D_1)\sum |F_{m,X}(u,v) - f_{m,X}(\xi)(v-u)| \\ &\quad + (D_1)\sum |f_{n,X}(\xi) - f_{m,X}(\xi)|(v-u) \\ &\quad + (D_2)\sum |F_{n,X}(u,v)| \\ &\quad + (D_2)\sum |F_{m,X}(u,v)|. \end{aligned}$$

Since each $f_{n,X}$ is Henstock integrable on $[a,b]$, we can choose a suitable δ-fine division D above so that

$$|F_{n,X}(x,y) - F_{m,X}(x,y)| < \varepsilon(4+b-a) \text{ for } n,m \ge N.$$

In other words, $F_X(x)$ exists for $x \in [a,b]$. In fact, for any partial division D of $[a,b]$ we have

$$|(D)\sum\{F_{n,X}(u,v) - F_{m,X}(u,v)\}| < \varepsilon \text{ for } n,m \ge N.$$

Applying the above result, we can find a subsequence $F_{n(j),X}$ of $F_{n,X}$ such that for any partial division D of $[a,b]$ we have

$$|(D)\sum\{F_{n(j),X}(u,v) - F_X(u,v)\}| < \varepsilon 2^{-j} \text{ for } j = 1,2,\ldots.$$

Then putting $M(\xi) = n(1)$, for infinitely many $m(\xi) = n(j) \geq M(\xi)$ we have

$$|(D)\sum F_{m(\xi),X}(u,v) - F_X(a,b)| \leq \sum_{j=1}^{\infty} \varepsilon 2^{-j} = \varepsilon.$$

The proof is complete.

Lemma 4. If the conditions in Lemma 3 are satisfied, then for every $\varepsilon > 0$ there is an integer N such that for every $n \geq N$ there is a $\delta(\xi) > 0$ such that for any δ-fine division D of X we have

$$(D)\sum F_{n,X}(u,v) - F_X(u,v)| < \varepsilon.$$

Proof. It follows from the first part of the proof of Lemma 3 by letting $m \to \infty$.

Lemma 5. Let f_n be Henstock integrable on $[a,b]$ with the primitive F_n for $n = 1,2,\ldots$. If F_n are $UAC^{**}(X)$ then for every $\varepsilon > 0$ there is a $\delta(\xi) > 0$, independent of n, such that for any δ-fine partial division D of X we have

$$|(D)\sum\{F_{n,X}(u,v) - F_n(u,v)\}| < \varepsilon.$$

Proof. Since F_n are $UAC^{**}(X)$, for every $\varepsilon > 0$ there are a $\delta(\xi) > 0$ and an $\eta > 0$, both independent of n, such that the rest of the condition for $AC^{**}(X)$ holds. For each n, there is a $\delta_n(\xi) > 0$ with $\delta_n(\xi) \leq \delta(\xi)$ such that for any δ_n-fine partial division D of $[a,b]$ we have

$$|(D)\sum\{F_{n,X}(u,v) - f_{n,X}(\xi)(v-u)\}| < \varepsilon,$$

$$|(D)\sum\{F_n(u,v) - f_n(\xi)(v-u)\}| < \varepsilon.$$

Here we have used the fact that $f_{n,X}$ is Henstock integrable on $[a,b]$ by Lemma 2. Now take any δ-fine partial division D of X and construct δ_n-fine division D_1 of the union of $[u,v]$ in D. Split D_1 into D_2 and D_3 so that D_2 contains the intervals with $\xi \in X$ and D_3 otherwise. Then we obtain

$$\begin{aligned}
|(D)\sum\{F_{n,X}(u,v) - F_n(u,v)\}| &= |(D_1)\sum\{F_{n,X}(u,v) - F_n(u,v)\}| \\
\leq |(D_2)\sum\{F_{n,X}(u,v) - F_n(u,v)\}| &+ |(D_3)\sum\{F_{n,X}(u,v) - F_n(u,v)\}| \\
&\leq |(D_2)\sum\{F_{n,X}(u,v) - f_{n,X}(\xi)(v-u)\}| \\
&+|(D_2)\sum\{f_n(\xi)(v-u) - F_n(u,v)\}| \\
&+|(D_3)\sum F_{n,X}(u,v)| + |(D_3)\sum F_n(u,v)|.
\end{aligned}$$

Modify $\delta(\xi)$ if necessary so that $(\xi-\delta(\xi),\xi+\delta(\xi))\subset(a,b)-X$ when $\xi\notin X$ and that $(D\backslash D_2)\sum|v-u|<\eta$ as in the proof of Lemma 2. Then $F_{n,X}(u,v)=0$ when $\xi\notin X$. Note that $D_3=D\setminus D_2$. Consequently,

$$|(D)\sum\{F_{n,X}(u,v)-F_n(u,v)\}|<3\varepsilon.$$

That is, the required condition holds.

Lemma 6. If each f_n is Henstock integrable on $[a,b]$ with $f_n(x)\to f(x)$ almost everywhere in $[a,b]$, the primitives F_n of f_n being $UACG^{**}$, then the condition in Lemma 1 holds.

Proof. Let $[a,b]=\cup_i X_i$ such that F_n are $UAC^{**}(X_i)$, for each i. In view of Lemma 5, when $X=X_i$ the functions $F_{n,X}$ are also $UAC^{**}(X)$. Then, by Lemma 4, for every $\varepsilon>0$ and X_i, writing $X=X_i$, there exist an integer $n=n(i,j)$ and a $\delta_n(\xi)>0$ such that for any δ_n-fine division D of X we have

$$(D)\sum|F_{n,X}(u,v)-F_X(u,v)|<\varepsilon 2^{-i-j}.$$

In view of lemma 5 again, we may modify $\delta_n(\xi)$ if necessary so that for the same D above we have

$$(D)\sum|F_{n,X}(u,v)-F_n(u,v)|<\varepsilon 2^{-i},$$

$$(D)\sum|F_X(u,v)-F(u,v)|<\varepsilon 2^{-i}.$$

Again, we have used Lemma 2 here. We may assume for each i that $F_{n(i,j)}$ is a subsequence of $F_{n(i-1,j)}$. Now consider $f_{n(j)}=f_{n(j,j)}$ in place of the original sequence.

Let $X_1=Y_1$ and $Y_i=X_i-(X_1\cup\ldots\cup X_{i-1})$ for $i=2,3,\ldots$. Given $\varepsilon>0$, take $M(\xi)=n(i)$ and $m(\xi)$ to have values in $\{n(j)\geq i\}$ when $\xi\in Y_i$. For such $m(\xi)$ put $\delta(\xi)=\delta_{m(\xi)}(\xi)$. Then for any δ-fine division D we have, writing $m=m(\xi)$ and $X=X_i$ when $\xi\in Y_i$,

$$|F_m(u,v)-F(u,v)|\leq|F_m(u,v)-F_{m,X}(u,v)|+|F_{m,X}(u,v)-F_X(u,v)|$$
$$+|F_X(u,v)-F(u,v)|$$

and consequently,

$$(D)\sum|F_{m(\xi)}(u,v)-F(u,v)|<3\varepsilon.$$

Hence the condition in Lemma 1 is satisfied.

Finally, we have completed the proof of Theorem 1.

4. THE REAL LINE

The situation is much simpler on the real line. Let f be Henstock integrable on $[a,b]$ with the primitive F. We can easily verify that F is ACG^{**}. In view of Lemma 2 and using the notation there, we obtain

$$F(u,v) = \int_u^v f_X(x)dx + \sum_{k=1}^{\infty} F([u,v] \cap [c_k, d_k])$$

where (c_k, d_k) are the intervals contiguous to X, assuming X is closed. Consequently, F is ACG^*. In other words, for the primitive F the conditions ACG^* and ACG^{**} are equivalent. Now we may restate Theorem 1 as follows.

Theorem 2. Let the following conditions be satisfied:

(i) $f_n(x) \to f(x)$ almost everywhere in $[a,b]$ as $n \to \infty$ where each f_n is Henstock integrable on $[a,b]$;

(ii) the primitives F_n of f_n are $UACG^*$;

(iii) the sequence F_n coverges uniformly to a limit function F on $[a,b]$.

Then f is also Henstock integrable on $[a,b]$ with the primitive F.

Proof. In view of (ii) and (iii), F is ACG^* and continuous. Since the Denjoy and Henstock integrals are equivalent, it remains to show that $F'(x) = f(x)$ almost everywhere.

Suppose F_n are $UAC^*(X)$ with X closed. If we can show that $F'(x) = f(x)$ almost everywhere in X, then the proof is complete in view of (ii). To do so, put $G_n(x) = F_n(x)$ when $x \in X$ and linear elsewhere in $[a,b]$. Then G_n are uniformly absolutely continuous on $[a,b]$. Writing $g_n(x) = G_n'(x)$ for almost all x, $g(x) = f(x)$ when $x \in X$ and $G'(x)$ elsewhere, we obtain by Vitali's convergence theorem or Lemma 3 with $X = [a,b]$ that g is Henstock integrable on $[a,b]$ with the primitive G. Furthermore, $f(x) = g(x) = G'(x) = F'(x)$ for almost all x in X. Hence we have proved the theorem.

In fact, (iii) may be replaced by: $F_n(x)$ converges pointwise to a continuous function $F(x)$ as $n \to \infty$ (see [14]). The same proof goes through. The proofs of the lemmas in Section 3 may also be simplified. For example, in Lemma 2 we may put

$$A = F(a,b) - \sum_{k=1}^{\infty} F(c_k, d_k)$$

where (c_k, d_k) are the intervals contiguous to X, assuming X is closed. Then we can show that f_X is Henstock integrable to A on $[a,b]$. Indeed, Lemma 2 holds with $AC^{**}(X)$ replaced by $VB^*(X)$. For a definition of $VB^*(X)$, see [18]. Lemma 3 follows from Vitali's convergence theorem. Lemma 5 is trivial since

$$|F_{n,X}(u,v) - F_n(u,v)| \le \sum_{k=1}^{\infty} |F_n([u,v] \cap [c_k, d_k])|$$

and the infinite series of the oscillations of F_n over $[c_k, d_k]$, $k = 1, 2, \ldots$, converges uniformly in n.

There have been some applications of the controlled convergence theorem. Liu [15] showed that in the definition of the Henstock integral we can always take measurable $\delta(\xi)$. The proof is straightforward if we use the controlled convergence theorem. Lee and Chew [11] used it to give a Riesz-type definition of the Henstock integral, i.e. every Henstock integrable function is the limit of a sequence of step functions satisfying the conditions in the controlled convergence theorem. Chew [3] extended the result to a nonlinear case and used it to prove a Riesz representation theorem for continuous additive functionals on the space of all Henstock integrable functions.

Here we give a proof of a theorem of Alexiewicz [1]. Let E denote the space of all Henstock integrable functions on $[a, b]$ provided with the norm

$$\|f\|_E = \sup\{|\int_a^x f(t)dt|;\ a \le x \le b\}.$$

This is a normed linear space though not complete. We define a continuous linear functional on E in the usual way.

Theorem 3. If T is a continuous linear functional on E, then

$$T(f) = \int_a^b f(x)g(x)dx$$

for all $f \in E$ and for some g of bounded variation on $[a, b]$.

Proof. First, we recall a result due to Riesz [16 page 75]. A function G is the primitive of a function of bounded variation on $[a, b]$ if and only if

$$\sum_{i=0}^{n-2} \left|\frac{G(x_{i+2}) - G(x_{i+1})}{x_{i+2} - x_{i+1}} - \frac{G(x_{i+1}) - G(x_i)}{x_{i+1} - x_i}\right| + \left|\frac{G(x_n) - G(x_{n-1})}{x_n - x_{n-1}}\right|$$

is bounded for all divisions $a = x_0 < x_1 < \ldots < x_n = b$. Now put $G(t) = T(\chi_{[a,t]})$ where $\chi_{[a,t]}$ denotes the characteristic function of $[a, t]$. We shall prove that G satisfies the above condition.

Writing

$$\psi_i = \frac{1}{x_{i+2} - x_{i+1}}\chi_{(x_{i+1},x_{i+2}]} - \frac{1}{x_{i+1} - x_i}\chi_{(x_i,x_{i+1}]}$$

when $i = 0, 1, \ldots, n-2$, and

$$= \frac{1}{x_n - x_{n-1}}\chi_{(x_{n-1},x_n]} \text{ when } i = n-1,$$

by the linearity and boundedness of T we obtain that the above expression is less than

$$\sum_{i=0}^{n-1} |T(\psi_i)| = T(\sum_{i=0}^{n-1} \epsilon_i\psi_i) \le M\|\sum_{i=0}^{n-1} \epsilon_i\psi_i\|_E \le 2M$$

where ϵ_i denotes $+1$ or -1 as the case may be. That is, the condition is satisfied and G is the primitive of a function say, g which is of bounded variation on $[a, b]$. Therefore the representation holds true for step functions.

Let $f \in E$. In view of a Riesz-type definition of the Henstock integrable functions [10 page 139], there is a sequence of step functions ψ_n converging to f and satisfying all the conditions in Theorem 2. Apply Theorem 2 and we have

$$\begin{aligned} T(f) &= \lim_{n\to\infty} T(\psi_n) \\ &= \lim_{n\to\infty} \int_a^b \psi_n(x)g(x)dx \\ &= \int_a^b f(x)g(x)dx. \end{aligned}$$

Hence the proof is complete.

As a bonus, we deduce the classical Riesz representation theorem as follows. Let C be the space of all continuous functions on $[a, b]$ and T a continuous linear funcitonal on C. Suppose F is continuous and ACG^*. Then F is the primitive of a Henstock integrable function f. By Theorem 3 there is a function g of bounded variation such that

$$T(F) = T^*(f) = \int_a^b fg = \int_a^b F dg_1.$$

Here T^* represent a continuous linear functional on E and the last step is by integration by parts in which g_1 is another function of bounded variation derived from g. Since every continuous function F is the limit of a uniformly convergent sequence of continuous functions $\{F_n\}$ which are also ACG^*, apply the uniform convergence theorem for the Stieltjes integral and we obtain

$$\begin{aligned} T(F) &= \lim_{n\to\infty} T(F_n) \\ &= \lim_{n\to\infty} \int_a^b F_n dg_1 \\ &= \int_a^b F dg_1. \end{aligned}$$

Hence T can be represented by means of the Stieltjes integral. Note that we have provided a proof of the classical Riesz representation theorem without resorting to the Hahn-Banach extension theorem.

5. THE n-DIMENSIONAL CASE

We shall indicate how the controlled convergence theorem and its proof may be extended to the n-dimensional case.

Let E be an interval in the n-dimensional space, that is, it is the set of all points $x = (x_1, \ldots, x_n)$ with $a_j \le x_j \le b_j$ for $j = 1, 2, \ldots, n$. Assume that a norm has been defined on the n-dimensional space. An open sphere $S(x, r)$ with centre x and radius r is the set of all y such that $\|x - y\| < r$. Let an interval E be given. A division D of E is a finite collection of interval-point pairs (I, x) with the intervals non-overlapping and their union E. It is said to be δ-fine if $I \subset S(x, \delta(x))$ where x is a vertex of I. Then a real number H is the value of the generalized Riemann integral of f over E if given $\varepsilon > 0$ there is a positive function $\delta(x)$ such that

$$|(D)\sum f(x)|I| - H| < \varepsilon$$

for all δ-fine division D of E. For more detail, see [7].

An elementary set is an interval or a finite union of non-overlapping intervals. Note that the difference set of two intervals is an elementary set. Let D_1 and D_2 be two partial divisions of X, i.e. the associated points lie in X. Let E_1 be the union of intervals in D_1 and E_2 the union of intervals in D_2. Then $D_1 \setminus D_2$ denotes the collection of component elementary sets (consisting of adjacent intervals) in $E_1 - E_2$. A function F defined on E is said to be $AC^{**}(X)$ with X being closed if for every $\varepsilon > 0$ there are a $\delta(x) > 0$ and an $\eta > 0$ such that for any two δ-fine partial divisions of X, D_1 and D_2, in which D_2 may be void, we have

$$(D_1 \setminus D_2) \sum |I| < \eta \text{ implying } (D_1 \setminus D_2) \sum |F(I)| < \varepsilon.$$

Note that I above denotes a typical component elementary set in $D_1 \setminus D_2$. Similarly, we define ACG^{**} and $UACG^{**}$.

We can verify that all the lemmas in Section 3 hold true for the n-dimensional case. Hence we have proved the following.

Theorem 4. Let the following conditions be satisfied:

(i) $f_n(x) \to f(x)$ everywhere in E as $n \to \infty$ where each f_n is generalized Riemann integrable on E;

(ii) the primitives F_n of f_n are $UACG^{**}$.

Then f is also generalized Riemann integrable on E and we have

$$\int_E f_n \to \int_E f \text{ as } n \to \infty.$$

Now the controlled convergence theorem has been proved for the n-dimensional case.

As pointed our by Chew, if f is generalized Riemann integrable on E with primitive F then it is easy to verify that F is $AC^{**}(X_i)$ with

$$X_i = \{x; |f(x)| \leq i\}.$$

Therefore F is ACG^{**} withour requiring X_i to be closed but only measurable. We remark that Theorem 3 still holds true using this new version of ACG^{**}. Indeed, the proof may proceed as follows. Suppose F_n are $UAC^{**}(X_i)$ for each i and $\cup_{i=1}^{\infty} X_i = E$. The there is a closed set $Y_i \subset X_i$ such that F_n are $UAC^* * (Y_i)$ and $S = E - \cup_{i=1}^{\infty} Y_i$ is of measure zero. In view of $UACG^{**}$, we have $|F(S)| = 0$. Hence the proof goes through as before. In other words, ACG^{**} in the new sense is a property of the primitive and not an additional condition imposed on it.

REFERENCES

1. A. Alexiewicz, Linear functionals on Denjoy integrable functions, Coll. Math. 1 (1948) 289-293.

2. V. G. Chelidze and A. G. Djvarsheishvili, Theory of the Denjoy integral and some of its applications (Russian), Tbilisi 1978.

3. T. S. Chew, Nonlinear Henstock-Kurzweil integrals and representation theorems, SEA Bull. Math. 12 (1988) 97-108.

4. A. G. Djvarsheishvili, On a sequence of integrals in the sense of Denjoy (Russian), Akad. Nauk Gruzin. SSR Trudy Mat. Inst. Rajmadze 18 (1951) 221-236.

5. R. Gordon, Equivalence of the generalized Riemann and restricted Denjoy integrals, Real Analysis Exchange 12 (1986-87) 551-574.

6. R. Henstock, Linear Analysis, London 1967.

7. R. Henstock, Lectures on the theory of integration, Singapore 1988.

8. Y. Kubota, A direct proof that the RC-integral is equivalent to the D^*-integral, Proc. Amer. Math. Soc. 80 (1980) 293-296.

9. P. Y. Lee, Lanzhou Lectures on Henstock integration, World Scientific 1989.

10. P. Y. Lee and T. S. Chew, A better convergance theorem for Henstock integrals, Bull. London Math. Soc. 17 (1985) 557-564.

11. P. Y. Lee and T. S. Chew, On convergence theorems for the nonabsolute integrals, Bull. Austral. Math. Soc. 34 (1986) 133-140.

12. P. Y. Lee and T. S. Chew, A short proof of the controlled convergence theorem for Henstock integrals, Bull. London Math. Soc. 19 (1987) 60-62.

13. P. Y. Lee and Wittaya Naak-In, A direct proof that Henstock and Denjoy integrals are equivalent, Bull. Malaysian Math. Soc. (2) 5 (1982) 43-47.

14. K. C. Liao, A refinement of the controlled convergence theorem for Henstock integrals, SEA Bull. Math. 11 (1987) 49-51.

15. G. Q. Liu, The measurability of $\delta(\xi)$ in Henstock integration, Real Analysis Exchange, 13 (1987/88) 446-450.

16. P. Muldowney, A general theory of integration in function spaces, Longman 1987.

17. F. Riesz and B. Sz.-Nagy, Functional Analysis (French), Budapest 1955.

18. S. Saks, Theory of the integral, 2nd ed., Warsaw 1937.

National University of Singapore
Republic of Singapore

On some aspects of open multifunctions

P. Maritz

1. INTRODUCTION

The hypothesis of openness of a multifunction occurs very rarely in the literature. This is perhaps due to the fact that such multifunctions are, under fairly general conditions, "almost constant". Franklin [14] was the first to obtain results of this nature. Image-open multifunctions were discussed briefly by Choquet [6]. More recently, Münnich and Száz [30] established a theorem which not only improves a result of Stanojević [38], but which has among its applications improvements of results of Ponomarev [34] and, in particular, of Franklin's results on constant and "almost constant" multifunctions.

Graph-open multifunctions seem to occur in a natural way in mathematical economics. Gale and Mas-Colell [15] proved the existence of a Walrasian General Equilibrium by means of irreflexive preference mappings (multifunctions, here) with open graphs. (The augmented preference mappings in their paper must not have open graphs.) Mas-Colell [25] proved that if a multifunction has an open graph and its values are homeomorphically convex (an open star-shaped set is homeomorphically convex), then it has a continuous selection. In applications it is not uncommon to encounter fixed point theory of graph-open multifunctions, see [15] p. 10. In [5] Ceder and Levi gave an example of a lower-semi-continuous graph-open multifunction with no continuous selection and they also showed that certain graph-open multifunctions may have Borel 1 selections. Shafer and Sonnenschein [37] proved the existence of equilibrium in an abstract economy with preferences which may be both non-transitive and non-complete by using "preference correspondences" (that is, multifunctions) with open graphs. In [26] and [27] McClendon proved fixed point and selection theorems for subopen multifunctions and also for multifunctions with r-open graphs. Henstock [16] and [17], Kurzweil [19] and Pfeffer [33], among others, employ positive functions in the integration theories that they consider. These positive functions are sometimes used in such a fashion that they may be replaced by image-open multifunctions; see the image-open multifunctions, called ***gauges***, in McLeod [28] and McShane [29].

In [28] the Riemann integrable functions are characterised as those generalised Riemann integrable functions for which there is a gauge γ on $[a, b]$ such that $\gamma(x) = (x - \delta,\, x + \delta)$, for some $\delta > 0$, and with the usual properties; such a gauge is lower-semi-continuous.

In some problems there is a great deal of freedom in defining a gauge to meet the requirements of the definition of the particular integral. In some cases a gauge γ with the property that $\gamma(x)$ is of constant width for all (or for almost all) x is sufficient; in other instances, a constant (or almost constant) gauge is chosen.

For example, if the function $f : [0, 1] \to \mathbb{R}$ is defined by

$$\begin{aligned} f(x) &= 3 \text{ if } x = 0 \\ &= 2 \text{ if } x \in (0, 1], \end{aligned}$$

then a suitable gauge is

$$\begin{aligned} F(x) &= (-\varepsilon, \varepsilon) \text{ if } x = 0 \\ &= (-1, 2) \text{ if } x \in (0, 1]. \end{aligned}$$

Here F is image-open and "almost constant". Furthermore, if the function $g : [0, 1] \to \mathbb{R}$ is defined by

$$\begin{aligned} g(x) &= 0 \text{ if } x \text{ is irrational in } [0, 1] \\ &= 1 \text{ if } x \text{ is rational in } [0, 1], \end{aligned}$$

then a suitable gauge for g is

$$\begin{aligned} G(x) &= (-2, 2) \text{ if } x \text{ is irrational in } [0, 1] \\ &= (r_k - \tfrac{\varepsilon}{2^{k+1}}, r_k + \tfrac{\varepsilon}{2^{k+1}}) \text{ if } x \text{ is rational in } [0, 1]. \end{aligned}$$

In this case G is "constant almost everywhere", that is, constant on the complement of a set of the first Baire category. This has the effect that we study constant quasi-continuous multifunctions in section 5.

The fact that familiar topological concepts defined in terms of open covers (such as paracompactness, for instance) can be reformulated in terms of image-open multifunctions, suggests that image-open multifunctions should be investigated thoroughly with their intrinsic properties and applications as primary objectives. See also Yusufov [41] for recent results on open covering mappings.

The purpose of this paper is thus to discuss properties of multifunctions possessing some notions of openness. Section 2 is reserved for notes regarding notation and preliminary results needed in later sections. In section 3 we consider some of the results of Münnich and Száz [30] together with counterexamples. Section 4, on quasi-connectedness, is designed in such a way that it provides the necessary background needed for an investigation of constant quasi-continuous multifunctions in section 5. Multifunctions with non-mingled values are being studied in section 6, together with two selection results on image-open and graph-open multifunctions.

2. NOTATIONS, DEFINITIONS AND PRELIMINARY RESULTS

Suppose X and Y are non-empty sets and $\mathcal{P}(Y)$ is the class of all subsets of Y. A function $F: X \to \mathcal{P}(Y)$ is called a multifunction, usually denoted by $F: X \to Y$; the set $\{x \in X | F(x) \neq \emptyset\}$ is called the effective domain of F. Following Bellenger and Simons [1], we call a multifunction strict if $F(x) \neq \emptyset$ for every $x \in X$. For $A \subset X$, let (α)

$$F(A) = \bigcup_{x \in A} F(x).$$

The set $F(X)$ is called the range of F; if $F(X) = Y$, then F maps X onto Y. For $B \subset Y$, let $F^+(B) = \{x \in X | F(x) \subset B\}$, the strong inverse of F, and let $F^-(B) = \{x \in X | F(x) \cap B \neq \emptyset\}$, the weak inverse of F. Then $F^-: Y \to \mathcal{P}(X)$ and if $y \in Y$, then $F^-(y) = \{x \in X | y \in F(x)\}$. Thus, for $B \subset Y$, we have (β)

$$F^-(B) = \bigcup_{y \in B} F^-(y).$$

Henceforth, X and Y will be fixed non-empty topological spaces and all multifunctions $F: X \to Y$ will be strict. The closure and the interior of a set $A \subset X$ will be denoted by $\bar{A}$ and A°, respectively. Let $\mathcal{O}(X)$ and $\mathcal{C}(X)$ be the classes of all open subsets and all closed subsets of X, respectively. All neighbourhoods of points are open. If $x \in X$, then $\mathcal{U}(x)$ denotes the neighbourhood system of x. A multifunction $F: X \to Y$ is image-P if $F(x)$ has property P for every $x \in X$. The boundary of a set $A \subset X$ is denoted by $B(A)$; thus $B(A) = \bar{A} \setminus A^\circ$.

Paracompactness of a space can be formulated in terms of image-open multifunctions. In fact, a space Y is paracompact if and only if it is Hausdorff and if for every image-open multifunction F from an index set Γ onto Y, there exists an image-open multifunction G from an index set Λ onto Y such that for every $\lambda \in \Lambda$ there exists a $\gamma_\lambda \in \Gamma$ satisfying $G(\lambda) \subset F(\gamma_\lambda)$ and such that for every $y \in Y$, there exists a set $U \in \mathcal{U}(y)$ with the property that $G^-(U)$ is a finite subset of Λ.

Following Neubrunn [32], we prefer to use the term quasi-open set instead of the original terminology of semi-open set as defined by Levine [20].

2.1 **Definition**. Let A be a subset of X.

1. ([20]) A is said to be quasi-open if there exists a set $U \in \mathcal{O}(X)$ such that $U \subset A \subset \bar{U}$.

2. ([7]) A is said to be quasi-closed if $X \setminus A$ is quasi-open.

Let $\mathcal{QO}(X)$ and $\mathcal{QC}(X)$ be the classes of all quasi-open and all quasi-closed subsets of X, respectively. Obviously, $\mathcal{O}(X) \subset \mathcal{QO}(X)$ and $\mathcal{C}(X) \subset \mathcal{QC}(X)$.

2.2 **Definition**. Let A be a subset of X.

1. ([4]) A is said to be a quasi-neighbourhood of a point $x \in X$ if there exists a set $B \in \mathcal{QO}(X)$ such that $x \in B \subset A$.

2. ([7]) The quasi-closure of A, denoted by $\underline{A}$, is defined by $\underline{A} = \cap\{C | C \in \mathcal{QC}(X) \text{ and } A \subset C\}$.

3. ([7]) The quasi-interior of A, denoted by $A_\circ$, is defined by $A_\circ = \cup\{O | O \in \mathcal{QO}(X) \text{ and } O \subset A\}$.

We suppose that all quasi-neighbourhoods of points are quasi-open. If $x \in X$, then $\mathcal{U}_q(x)$ denotes the quasi-neighbourhood system of x.

2.3 **Lemma**. Let A be a subset of X.

1. ([7]) $A^\circ \subset A_\circ \subset A \subset \underline{A} \subset \bar{A}$.

2. ([8]) If A is quasi-open (quasi-closed), then A°, $A_\circ$, $\underline{A}$ and $\bar{A}$ are quasi-open (quasi-closed).

3. ([20]) If $A \in \mathcal{QO}(X)$ and if $A \subset B \subset \bar{A}$, then $B \in \mathcal{QO}(X)$.

The graph of a multifunction $F: X \to Y$ is the set $G(F) = \{(x, y) \in X \times Y | y \in F(x)\}$.

2.4 **Definition**. Let $F: X \to Y$ be a multifunction.

1. F is said to be open if $F(A) \in \mathcal{O}(Y)$ for every set $A \in \mathcal{O}(X)$.

2. ([3]) F is said to be quasi-open if $F(A) \in \mathcal{QO}(Y)$ for every set $A \in \mathcal{O}(X)$.

3. ([9]) F is said to be pre-quasi-open if $F(A) \in \mathcal{QO}(Y)$ for every set $A \in \mathcal{QO}(X)$.

4. F is said to be graph-open (graph-quasi-open) if

$$G(F) \in \mathcal{O}(X \times Y) \;\; (G(F) \in \mathcal{QO}(X \times Y)).$$

Clearly, if F is open or pre-quasi-open, then F is quasi-open.

2.5 **Lemma**. Let $F: X \to Y$ be a multifunction.

1. If F is image-open, then F is open, quasi-open and pre-quasi-open.

2. If F is image-quasi-open, then F is both quasi-open and pre-quasi-open.

Proof. Recall that every open set is quasi-open and that an arbitrary union of quasi-open sets is quasi-open, see [20] Theorem 2. Then apply (α).

The following example shows that the converse of each of 2.5(1) and (2) is false.

2.6 **Example**. Equip $\mathbb{R}$ (the set of real numbers) with its usual topology and let $F: \mathbb{R} \to \mathbb{R}$ be defined by $F(x) = \{x\}$ for every $x \in \mathbb{R}$. Then F is open, quasi-open and pre-quasi-open, but not image-quasi-open (and hence not image-open).

2.7 **Lemma**. Let A and B be subsets of X.

1. ([7]) If $A \in \mathcal{QO}(X)$ and $B \in \mathcal{O}(X)$, then $A \cap B \in \mathcal{QO}(X)$.

2. ([20]) $A \in \mathcal{QO}(X)$ if and only if $A \subset \overline{(A^\circ)}$.

2.8 **Definition ([18], [24])**. A function $f: X \to Y$ is said to be quasi-continuous at a point $a \in X$ if for every $U \in \mathcal{U}(a)$ and for every $W \in \mathcal{U}(f(a))$, there is a non-empty open set $G \subset U \cap f^{-1}(W)$; f is quasi-continuous on X if it is quasi-continuous at every point of X.

2.9 **Lemma**. A function $f: X \to Y$ is quasi-continuous on X if and only if $f^{-1}(V) \in \mathcal{QO}(X)$ for every set $V \in \mathcal{O}(Y)$.

Proof. Let f be quasi-continuous on X and let $V \in \mathcal{O}(Y)$. If $f^{-1}(V) = \emptyset$, then $f^{-1}(V) \in \mathcal{QO}(X)$. If $f^{-1}(V) \neq \emptyset$, let $a \in f^{-1}(V)$ and let $U \in \mathcal{U}(a)$. There is a non-empty set $G \subset U \cap f^{-1}(V)$. Then $G = G^\circ \subset (f^{-1}(V))^\circ$, and so $U \cap (f^{-1}(V))^\circ \neq \emptyset$. This implies that $a \in \overline{(f^{-1}(V))^\circ}$, thus $f^{-1}(V) \subset \overline{(f^{-1}(V))^\circ}$. Hence $f^{-1}(V) \in \mathcal{QO}(X)$ by 2.7(2). Conversely, let $f^{-1}(V) \in \mathcal{QO}(X)$ for every set $V \in \mathcal{O}(Y)$. Let $a \in X$, $U \in \mathcal{U}(a)$ and $V \in \mathcal{U}(f(a))$. Then $\emptyset \neq U \cap f^{-1}(V) \in \mathcal{QO}(X)$, by 2.7(1). By 2.1(1), there exists a non-empty set $G \in \mathcal{O}(X)$ such that $G \subset U \cap f^{-1}(V)$. Then f is quasi-continuous at a, and hence also on X.

2.10 **Definition**. Let $F: X \to Y$ be a multifunction.

1. ([2]) F is said to be upper-semi-continuous, briefly u-s-c (lower-semi-continuous, briefly, l-s-c) at a point $a \in X$ if for every set $V \in \mathcal{O}(Y)$ satisfying $F(a) \subset V$ $(F(a) \cap V \neq \emptyset)$, there exists a set $U \in \mathcal{U}(a)$ such that $U \subset F^+(V)$ $(U \subset F^-(V))$. F is u-s-c (l-s-c) on X if it is u-s-c (l-s-c) at every point of X and F is semi-continuous at a point $a \in X$ (on X) if it is both u-s-c and l-s-c at the point a (on X).

2. ([18], [35]) F is said to be upper-quasi-continuous, briefly u-q-c (lower-quasi-continuous, briefly l-q-c) at a point $a \in X$ if for every set $V \in \mathcal{O}(Y)$ satisfying $F(a) \subset V$ $(F(a) \cap V \neq \emptyset)$ and for every set $U \in \mathcal{U}(a)$, there exists a set $G \in \mathcal{O}(X)$, $G \subset U$, $G \neq \emptyset$ such that $G \subset F^+(V)(G \subset F^-(V))$. F is u-q-c (l-q-c) on X if it is u-q-c (l-q-c) at every point of X and F is quasi-continuous at a point $a \in X$ (on X) if it is both u-q-c and l-q-c at the point a (on X).

2.11 **Lemma**. Let $F: X \to Y$ be a multifunction.

1. ([2]) F is u-s-c (l-s-c) on X if and only if $F^+(V) \in \mathcal{O}(X)$ $(F^-(V) \in \mathcal{O}(X))$ for every set $V \in \mathcal{O}(Y)$.

2. ([23], [32]) F is u-q-c (l-q-c) on X if and only if $F^+(V) \in \mathcal{QO}(X)$ ($F^-(V) \in \mathcal{QO}(X)$) for every set $V \in \mathcal{O}(Y)$.

It is clear that if $F: X \to Y$ is semi-continuous (upper, lower), then it is quasi-continuous (upper, lower), but the converse is not true.

2.12 **Example**. Let $X = \{a, b, c, d\}$ and equip X with the topology

$$\mathcal{O}(X) = \{\emptyset, \{a\}, \{b, c\}, \{a, b, c\}, X\}.$$

Then

$$\mathcal{C}(X) = \{\emptyset, \{d\}, \{a, d\}, \{b, c, d\}, X\},$$

$$\mathcal{QO}(X) = \{\emptyset, \{a\}, \{b, c\}, \{a, d\}, \{a, b, c\}, \{b, c, d\}, X\}$$

and

$$\mathcal{QC}(X) = \{\emptyset, \{a\}, \{d\}, \{a, d\}, \{b, c\}, \{b, c, d\}, X\}.$$

Define $F: X \to X$ as follows: $F(a) = \{b, c\}$, $F(b) = \{a, d\}$, $F(c) = X$ and $F(d) = \{b, c\}$. Then F is u-q-c, but none of l-q-c, u-s-c or l-s-c. Define $G: X \to X$ as follows: $G(a) = G(d) = \{a\}$ and $G(b) = G(c) = \{b, c\}$. Then G is quasi-continuous, but neither u-s-c nor l-s-c.

2.13 **Lemma**. Let $F: X \to Y$ be a multifunction.

1. If F^- is image-open, then F is l-s-c and l-q-c.

2. If F^- is image-quasi-open, then F is l-q-c.

Proof. This follows from (β).

That the converse of each of 2.13(1) and (2) is false can be seen from 2.6 where F is l-s-c and l-q-c, but F^- is not image-quasi-open (and hence not image-open).

2.14 **Lemma**. Let $F: X \to Y$ be a multifunction and let B be any subset of Y.

1. ([30]) If F is image-open and u-s-c, then $F^-(B) \in \mathcal{C}(X)$; in particular, F^- is image-closed.

2. If F is image-open and u-q-c, then $F^-(B) \in \mathcal{QC}(X)$; in particular, F^- is image-quasi-closed.

Proof. (2) Let $a \in X \setminus F^-(B) = F^+(Y \setminus B)$. Then $F(a) \subset Y \setminus B$. Since $F(a) \in \mathcal{O}(Y)$ and F is u-q-c, it follows that $F^+(F(a)) \in \mathcal{QO}(X)$, and hence $F^+(F(a)) \in \mathcal{U}_q(a)$. Since $a \in F^+(F(a)) \subset F^+(Y \setminus B) = X \setminus F^-(B)$, it follows that $a \in (X \setminus F^-(B))_\circ$; hence $X \setminus F^-(B) \in \mathcal{QO}(X)$. Consequently, $F^-(B) \in \mathcal{QC}(X)$.

2.15 **Lemma ([30])**. Let $F: X \to Y$ be a multifunction. If F is graph-open, then both F and F^- are image-open, hence F is l-s-c.

Example 1.6 in [30] shows that the converse of 2.15 above is false. The next example shows that if F is graph-quasi-open, then neither F nor F^- need be image-quasi-open. See also 5.5.

2.16 **Example**. Equip $\mathbb{R}$ with its usual topology and let $F : \mathbb{R} \to \mathbb{R}$ be defined by

$$\begin{aligned} F(x) &= \ ]1,\, 2[\text{ if } x < 2 \\ &= [1,\, 2] \text{ if } x = 2 \\ &= \ ]2,\, 3[\text{ if } x > 2. \end{aligned}$$

Then

$$\begin{aligned} F^-(y) &= \emptyset \text{ if } y \geq 3 \text{ or } y < 1 \\ &= \ ]2,\, \infty[\text{ if } y \in]2,\, 3[\\ &= \{2\} \text{ if } y \in \{1,\, 2\} \\ &= \ ]-\infty,\, 2[\text{ if } y \in]1,\, 2[. \end{aligned}$$

It is easily seen that $G(F) \in \mathcal{QO}(X \times Y)$, although neither F nor F^- is image-quasi-open.

The reason for the failure of the analogue of 2.15 for graph-quasi-open F is because if $T = T_1 \times T_2$, where T_1 and T_2 are topological spaces, and $A \in \mathcal{QO}(T)$, then A is in general not a union of sets of the form $A_1 \times A_2$, where $A_1 \in \mathcal{QO}(T_1)$ and $A_2 \in \mathcal{QO}(T_2)$.

In the light of 2.15 and [30] Example 1.6, Münnich and Száz posed the following question:

2.17 **Question**. When does the image openness of F or/and F^- imply that F is graph-open?

See [30] Corollary 2.6 and also 3.9 of this paper for a partial answer to this question. The last result of this section is actually [13] Theorem 7(1), adapted to the present situation, and it should be read in conjuction with 6.4, 6.10 - 6.12 of the present paper.

2.18 **Theorem**. Let Y be regular and $F : X \to Y$ be pre-quasi-open, u-q-c and image-co-dense. Then F is l-q-c.

Proof. Suppose that F is not l-q-c. Then there is a set $G \in \mathcal{O}(Y)$ such that $F^-(G) \notin \mathcal{QO}(X)$. Then $\emptyset \neq F^-(G) \neq X$, and by 2.7(2) there is a point $a \in F^-(G) \setminus (\overline{F^-(G)})^\circ$. Then $a \in B(F^-(G))$. Let $b \in F(a) \cap G$. Since Y is regular, there exists a set $V \in \mathcal{O}(Y)$ such that $b \in V \subset \bar{V} \subset G$. Then $a \in F^-(\bar{V}) \subset F^-(G)$, hence $a \in B(F^-(\bar{V}))$. Since F is u-q-c, $F^+(Y \setminus \bar{V}) = X \setminus F^-(\bar{V}) \in \mathcal{QO}(X)$. Since $\emptyset \neq F^-(\bar{V}) \subset F^-(G) \neq X$, it follows that $\emptyset \neq F^+(Y \setminus \bar{V}) \neq X$. Now, $a \in B(F^-(\bar{V})) = B(X \setminus F^-(\bar{V})) = B(F^+(Y \setminus \bar{V}))$. Since $\{a\} \subset \overline{F^+(Y \setminus \bar{V})}$ and $F^+(Y \setminus \bar{V}) \subset \overline{F^+(Y \setminus \bar{V})}$, it follows that

$$F^+(Y \setminus \bar{V}) \subset \{a\} \cup F^+(Y \setminus \bar{V}) \subset \overline{F^+(Y \setminus \bar{V}}.$$

Let $U = \{a\} \cup F^+(Y \setminus \bar{V})$. Then $U \in \mathcal{QO}(X)$ by 2.3(3). By the fact that F is pre-quasi-open, it follows that $F(U) \in \mathcal{QO}(Y)$ and that $F(U) \cap V =$

$F[\{a\} \cup F^+(Y \setminus \bar{V})] \cap V = F(a) \cap V \in \mathcal{QO}(Y)$ by 2.7(1). Then $(F(U) \cap V)^\circ = (F(a) \cap V)^\circ \subset (F(a))^\circ = \emptyset$, since F is image-co-dense. On the other hand, $b \in F(a) \cap V$, thus $(F(a) \cap V)^\circ \neq \emptyset$ by 2.1(1). We conclude that F is l-q-c.

3. CONSTANT SEMI-CONTINUOUS MULTIFUNCTIONS

In this section we shall, for the sake of completeness and to provide some counterexamples, discuss the alternative theorem (and some of its corollaries) of Münnich and Száz [30]. This theorem improves Lemma 3.5 of Stanojević [38] and it can be used to obtain not only an improvement of Theorem 4.1 of Ponomarev [34] (see [38]), but also to improve [14] Proposition 1 and to obtain [14] Proposition 2. This alternative theorem also gives a partial answer to 2.17.

3.1 **Theorem ([30])**. Let X be connected, $F : X \to Y$ be semi-continuous on X and let V be a subset of Y such that at least one of the following conditions is fulfilled:

1. V is clopen.

2. F is image-open and V is closed.

3. F^- is image-open and V is open.

4. Both F and F^- are image-open.

Then either $F^+(V) = X$ or $F^-(Y \setminus V) = X$.

The following example shows that 3.1 fails if F is not semi-continuous.

3.2 **Example**. Let $X = \{a, b, c\}$ with topology $\mathcal{O}(X) = \{\emptyset, \{a\}, \{a, b\}, X\}$ and let $Y = \{1, 2, 3\}$ with topology $\mathcal{O}(Y) = \{\emptyset, \{1\}, \{2, 3\}, Y\}$. Then X is a connected space, but Y is not. Define $F : X \to Y$ as follows: $F(a) = F(b) = Y$ and $F(c) = \{2, 3\}$. Then F and F^- are both image-open. If $V = \{2, 3\}$, then V is clopen in Y and $F^+(V) = \{c\}$ and $F^-(Y \setminus V) = \{a, b\}$. Notice that F is not semi-continuous.

3.3 **Corollary ([30])**. Let X be connected and $F : X \to Y$ be semi-continuous and onto Y such that $F(a)$ is connected in Y for some $a \in X$. Then Y is also connected.

The following example shows that the hypotheses of 3.3 cannot be weakened.

3.4 **Example**.

1. Let $X = \mathbb{R}$ with its usual topology, $Y = \mathbb{R}$ with the right half-open interval topology, [39] p.75, and let $F : X \to Y$ be defined by $F(x) = Y$ if $x \in X \setminus \{1\}$ and $F(1) = \{1\}$. Then all the conditions of 3.3, except the semi-continuity of F, are satisfied. However, Y is disconnected.

2. Let $X = \mathbb{R}$ with the right half-open interval topology, $Y = \{0, 1\}$ with the discrete topology and let $F : X \to Y$ be defined by $F(x) = \{1\}$ if $x \geq 1$ and $F(x) = \{0\}$ if $x < 1$. Then all the conditions of 3.3, except the connectedness of X, are satisfied. Again, Y is disconnected.

3.5 **Corollary ([30]).** Let X be connected and $F: X \to Y$ be semi-continuous and onto Y such that $F(a) \subset V$ for some $a \in X$, where V is a component of Y. Suppose that either F is image-open or V is open. Then Y is also connected.

The following example shows that 3.5 fails if F is not semi-continuous.

3.6 **Example.** Let X, $\mathcal{O}(X)$, Y, $\mathcal{O}(Y)$, F and V be as in 3.2. Then $V = F(c)$ is a component of Y, but Y is disconnected.

3.7 **Corollary ([30]).** Let $F: X \to Y$ be image-open and semi-continuous such that either F is image-closed or F^- is image-open. Then F is constant on each component of X.

Proposition 1 of Franklin [14] states that if $F: X \to Y$ is graph-open and u-s-c on X, then F is constant on each component of X. Lemma 2.15 shows that 3.7 is indeed an improvement of Franklin's Proposition 1. Example 3.8(1) below shows that 3.7 and Franklin's Proposition 1 fail if F is not semi-continuous on X and 3.8(2) shows that Franklin's Proposition 1 has no analogue for graph-quasi-open and quasi-continuous multifunctions.

3.8 **Example.**

1. Let X, $\mathcal{O}(X)$, Y, $\mathcal{O}(Y)$ and F be as in 3.2. Then X is connected, F is image-clopen and graph-open, F^- is image-open, but, of course, F is not constant on X.

2. Let $X = [1, 2]$ and $Y = \mathbb{R}$, both equipped with their usual topologies. Let $F: X \to Y$ be defined by

$$F(x) = \,]1, 2[\text{ if } x \in \,]1, 2[$$
$$= [1, 2] \text{ if } x \in \{1, 2\}.$$

Then X is connected, graph-quasi-open and quasi-continuous on X, but F is not constant on X.

The next corollary gives a partial answer to 2.17.

3.9 **Corollary ([30]).** Let X, Y and F be as in 3.7, and suppose that the components of X are open. Then F is graph-open.

Examples of spaces whose components are open are locally connected spaces and spaces whose classes of components are locally finite. The next result is exactly [14] Proposition 2.

3.10 **Corollary ([30]).** Let $F: X \to Y$ be image-open and semi-continuous. Then the multifunction $T: X \to Y$, defined by $T(x) = \overline{F(x)}$ for every $x \in X$, is constant on each component of X.

Proof. Let X be connected and let $V = T(a)$ for some $a \in X$. By 3.1(2), $F(X) \subset T(a)$. Then $\overline{F(X)} \subset T(a)$. Furthermore, (γ)

$$\overline{F(X)} = \overline{\bigcup_{x \in X} F(x)} \supset \bigcup_{x \in X} \overline{F(x)} = \bigcup_{x \in X} T(x) = T(X) \supset T(a).$$

Consequently, $T(a) = \overline{F(X)}$ for every $a \in X$.

3.11 **Example.**

1. The multifunction F in 3.2 shows that 3.10 fails if F is not semi-continuous.

2. Let $X = \{a, b, c, d\}$ with topology

$$\mathcal{O}(X) = \{\emptyset, \{a\}, \{a, b\}, \{a, c\}, \{a, d\}, \{a, b, c\}, \{a, c, d\}, \{a, b, d\}, X\}$$

and let $Y = \{1, 2, 3, 4\}$ with topology $\mathcal{O}(Y) = \{\emptyset, \{1, 3\}, Y\}$. Then X is a connected space, and Y is not T_1, see 3.12 and 3.13(2). Define $F: X \to Y$ by $F(a) = \{1, 3\}$, $F(b) = F(c) = F(d) = Y$ Then F satisfies the conditions of 3.10, and if $T(x) = \overline{F(x)}$ for every $x \in X$, then T is constant on X. If however,

$$\mathcal{O}_1(Y) = \{\emptyset, \{1,\}, \{2, 3\}, \{1, 2, 3\}, Y\}$$

and $G: X \to Y$, Y equipped with $\mathcal{O}_1(Y)$, is defined by $G(a) = G(c) = G(d) = \{1, 2, 3\}$ and $G(b) = \{1, 4\}$, then G is semi-continuous, not image-open, and if $T(x) = \overline{G(x)}$ for every $x \in X$, then T is not constant on X.

The next result is a combination of [30] Corollaries 2.8, 2.10; although the space Y in [30] Corollary 2.10 was required to be regular, we can combine these corollaries under the assumption that Y is T_1, and note that the said Corollary 2.10 also holds if Y is regular (without being T_1).

3.12 **Corollary ([30]).** Let $F: X \to Y$ be u-s-c, F^- image-open and Y a T_1-space.

1. Then F is constant on each component of X.

2. Then the multifunction $T: X \to Y$, defined by $T(x) = \overline{F(x)}$ for every $x \in X$, is constant on each component of X. (This result is also true if Y is regular, but not T_1).

Proof. Since the restrictions of F to the components of X have the same properties as F , we may assume that X is connected. By 2.13(1), F is l-s-c; thus F is semi-continuous and so 3.1 may be applied. Let $a \in X$.

1. Let $\mathcal{V} = \{V \in \mathcal{O}(Y) | F(a) \subset V\}$. By 3.1(3), and since $F(a) \subset V$ for every set $V \in \mathcal{V}$, we deduce that $F(X) \subset \cap\mathcal{V}$; thus $F(a) \subset F(X) \subset \cap\mathcal{V}$. Since Y is T_1, we have that $\cap\mathcal{V} \subset F(a)$. Consequently, $F(a) = F(X)$ for any $a \in X$.

2. Let $\mathcal{U} = \{U \in \mathcal{O}(Y) | T(a) \subset U\}$. By 3.1(3), and since $F(a) \subset U$ for every set $U \in \mathcal{U}$, we deduce that $F(X) \subset \cap\mathcal{U}$. Since Y is T_1, we have that $\cap\mathcal{U} \subset T(a)$. Then $\overline{F(X)} \subset T(a)$. By (γ) in 3.10, we have that $\overline{F(X)} \supset T(a)$. Consequently, $T(a) = \overline{F(X)}$ for any $a \in X$.

The following example shows that the hypotheses of 3.12 cannot be weakened.

3.13 **Example.**

1. Let X, Y and F be as in 3.4(1). Then X is connected, Y is T_1, F^- is image-open, but F is not u-s-c and neither F nor T is constant on X.

2. Let X, $\mathcal{O}(X)$, Y and $\mathcal{O}(Y)$ be as in 3.11(2). Define $F: X \to Y$ by $F(a) = F(b) = Y$, $F(c) = F(d) = \{2, 4\}$. Then F is u-s-c and F^- is image-open, but neither F nor T is constant on X.

4. QUASI-CONNECTEDNESS

This section is devoted to a short discussion of quasi-connected sets as defined by Das [10] and Dorsett [11].

4.1 **Definition ([10]).**

1. Two non-empty subsets A and B of X are said to be quasi-separated if $A \cap \underline{B} = \emptyset = \underline{A} \cap B$.

2. A subset A of X which cannot be expressed as the union of two quasi-separated sets is said to be quasi-connected.

It follows from 2.3(1) that if A and B are separated, then they are quasi-separated. Simple examples show that the converse is not true. Every quasi-connected set is connected. To see this, let X be quasi-connected and let A and B be separated subsets of X. Since A and B are quasi-separated, it follows that $X \neq A \cup B$. Hence, X is connected. The converse is not true, because, for example, $\mathbb{R}$ is connected, while $\mathbb{R} = (-\infty, 5] \cup (5, \infty)$ and the sets $(-\infty, 5]$ and $(5, +\infty)$ are quasi-separated; hence $\mathbb{R}$ is not quasi-connected.

In correspondence to connected spaces we have the following three results. The proofs of 4.3 and 4.4 are as in [12], adapted to the present situation.

4.2 **Lemma.** For every topological space X the following conditions are equivalent:

1. X is quasi-connected.

2. $\emptyset$ and X are the only quasi-open and quasi-closed (i.e., quasi-clopen) subsets of X.

3. If $X = A \cup B$, and the sets A and B are quasi-separated, then one of them is empty.

Proof. (1) $\Rightarrow$ (2): Suppose A is a non-empty quasi-clopen proper subset of X. Let $B = X \setminus A$. Then B is also a non-empty quasi-clopen proper subset of X. Since $X = A \cup B$ and A and B are quasi-separated, we have a contradiction. Thus, $A = \emptyset$ or $A = X$. (2) $\Rightarrow$ (3): If $X = A \cup B$ and $A \cap \underline{B} = \emptyset = \underline{A} \cap B$, then $\underline{A} \subset X \setminus B \subset A$ and $\underline{B} \subset X \setminus A \subset B$, hence $A, B \in \mathcal{QC}(X)$. Since $B = X \setminus A$

and $A = X \setminus B$, it follows that $A, B \in \mathcal{QO}(X)$. By (2), either $A = \emptyset$ or $B = \emptyset$.
(3) $\Rightarrow$ (1): Trivial.

4.3 **Lemma**. A subset C of X is quasi-connected if and only if for every pair A, B of quasi-separated subsets of X such that $C = A \cup B$, we have either $A = \emptyset$ or $B = \emptyset$.

4.4 **Lemma**. If a subset C of X is quasi-connected, then for every pair A, B of quasi-separated subsets of X such that $C \subset A \cup B$, we have either $C \subset A$ or $C \subset B$.

4.5 **Lemma**. Let $\{C_\alpha | \alpha \in \Lambda\}$ be a class of quasi-connected subsets of X. If there exists an $\alpha_o \in \Lambda$ such that C_{α_o} is not quasi-separated from any of the sets $C_{\alpha'}$ then $\bigcup_{\alpha \in \Lambda} C_\alpha$ is quasi-connected.

Proof. Let $C = \bigcup_{\alpha \in \Lambda} C_\alpha = A \cup B$, where A and B are quasi-separated. By 4.4, $C_{\alpha_o} \subset A$ or $C_{\alpha_o} \subset B$. Suppose $C_{\alpha_o} \subset A$. Then $C_\alpha \subset A$ for every $a \in \Lambda$ since no C_α is separated from C_{α_o}; thus $C \subset A$. Then $B = \emptyset$ and C is quasi-connected by 4.3.

4.6 **Lemma**. If the class $\{C_\alpha | \alpha \in \Lambda\}$ of quasi-connected subsets of X has non-empty intersection, then the set $\bigcup_{\alpha \in \Lambda} C_\alpha$ is quasi-connected.

Proof. If $\bigcap_{\alpha \in \Lambda} C_\alpha \neq \emptyset$, then no set C_α is quasi-separated from another. By 4.5, $\bigcup_{\alpha \in \Lambda} C_\alpha$ is quasi-connected.

4.7 **Lemma**. If a subset C of X is quasi-connected, then every subset A of X which satisfies $C \subset A \subset \underline{C}$ is quasi-connected.

Proof. Apply 4.5 to the class $\{C \cup \{x\} | x \in A\}$ with $C = C_{\alpha_o}$.

4.8 **Corollary**. If C is a quasi-connected subset of X, then $\underline{C}$ is quasi-connected.

4.9 **Remark**. The closure of a quasi-connected set is not necessarily quasi-connected. To see this, let $A = \{b, c\}$ in 2.12. Then A is quasi-connected but $\bar{A} = \{b, c, d\}$ is not quasi-connected. Note also that the union of the quasi-closed sets $\{a\}$ and $\{b, c\}$ is not quasi-closed.

The concept of a quasi-component of a point $x \in X$ defined below is based on the concept of a quasi-component of X as defined by Das [10] and Dorsett [11].

4.10 **Definition**. A quasi-connected subset A of X is said to be the quasi-component of a point $x \in X$ if $x \in A$ and if A is not a proper subset of a quasi-connected subset of X.

4.11 **Remark**.

1. Every $x \in X$ has a quasi-component, it may even be just $\{x\}$.

2. By 4.8, the quasi-component of a point is quasi-closed.

3. Suppose that $x, y \in X$, $x \neq y$ and that A_x and A_y are the quasi-components of x and y, respectively. If $A_x \neq A_y$ and $A_x \cap A_y \neq \emptyset$, then by 4.6, the set $A = A_x \cup A_y$ is quasi-connected. Then $A_x \subset \underline{A}$, $A_x \neq \underline{A}$, which contradicts the fact that A_x is a quasi-component. The quasi-components of two disjoint points of X either coincide or are disjoint,

so that the class of quasi-components constitutes a decomposition of X into pairwise disjoint quasi-closed and quasi-connected subsets, called the quasi-components of X.

Quasi-component as defined in 4.10 differs from the concept with the same name as defined by Engelking [12] p.438.

4.12 **Lemma**. A space X is quasi-connected if and only if there does not exist two non-empty disjoint quasi-open subsets of X.

Proof. Let X be quasi-connected . Suppose that A and B are non-empty disjoint quasi-open subsets of X. Then there is a non-empty open subset U of A and $X = \bar{U} \cup (X \setminus \bar{U})$. It follows directly from 2.3(2) that

$$\underline{(\bar{U})} \cap (X \setminus \bar{U}) = \bar{U} \cap (X \setminus \bar{U}) = \emptyset = \bar{U} \cap \underline{(X \setminus \bar{U})},$$

which implies that X is not quasi-connected, a contradiction. Let there be no two non-empty disjoint quasi-open subsets of X. Suppose, however, that X is not quasi-connected. By 4.2, there exists a non-empty quasi-clopen proper subset of X, which is a contradiction.

4.13 **Definition ([22])**. X is said to be quasi-T_2 if to each pair of distinct points x, $y \in X$, there exist disjoint sets $U \in \mathcal{U}_q(x)$ and $V \in \mathcal{U}_q(y)$.

Suppose that X contains at least two points. It is evident from 4.12 and 4.13 that if X is quasi-connected, then it is not quasi-T_2, and if X is quasi-T_2, then it is not quasi-connected.

For the next result we require a notion of denseness of classes of subsets of a topological space.

4.14 **Definition ([31])**. A class $\mathcal{A} \subset \mathcal{P}(X)$ is said to be upper-dense in a class $\mathcal{B} \subset \mathcal{P}(X)$, if for any set $B \in \mathcal{B}$ and for any open set $G \subset B$, there is a set $A \in \mathcal{A}$ such that $A \subset G$. A class $\mathcal{A}$ is said to be lower-dense in a class $\mathcal{B}$, if for any set $B \in \mathcal{B}$, any $y \in B$ and any $V \in \mathcal{U}(y)$, there is a set $A \in \mathcal{A}$ such that $A \cap V \neq \emptyset$.

Part 1 of the next result generalises [11]Theorem 3.1.

4.15 **Lemma**. Let X be quasi-connected and $F: X \to Y$ be u-q-c onto Y such that $\{F(x)|x \in X\}$ is upper-dense in $\mathcal{O}(Y)$.

1. Then Y is quasi-connected.

2. If Y is quasi-T_2, then F is constant on X.

Proof.

1. Suppose that Y is not quasi-connected. By 4.12 there exist two non-empty disjoint open sets U and V in Y. Then $F^+(U)$ and $F^+(V)$ are non-empty disjoint quasi-open sets in X, which imply that X is not quasi-connected, a contradiction.

2. If x, $y \in Y = F(X)$ and $x \neq y$, then Y is not quasi-connected. Thus F(X) is a singleton.

5. CONSTANT QUASI-CONTINUOUS MULTIFUNCTIONS

In this section we discuss the analogues of the results of section 3 for quasi-continuous multifunctions. Some of the results of section 3 are special cases of results of this section. Some counterexamples are given.

5.1 **Theorem**. Let X be quasi-connected, $F: X \to Y$ be quasi-continuous on X and let V be a subset of Y such that at least one of the following conditions is fulfilled:

1. V is clopen.
2. F is image-open and V is closed.
3. F^- is image-quasi-open and V is open.
4. F is image-open and F^- is image-quasi-open.

Then either $F^+(V) = X$ or $F^-(Y \setminus V) = X$.

Proof.

1. Let V be clopen. Then $F^+(V) \in \mathcal{QO}(X)$ because F is u-q-c and $F^+(V) \in \mathcal{QC}(X)$ because F is l-q-c. By 4.2(2), $F^+(V) = X$ or $F^+(V) = \emptyset$. Hence, $F^-(Y \setminus V) = \emptyset$ or $F^-(Y \setminus V) = X$.
2. Let F be image-open and V be closed. Then $F^-(Y \setminus V) \in \mathcal{QO}(X)$ because F is l-q-c. By 2.14(2), $F^-(Y \setminus V) \in \mathcal{QC}(X)$. Then $F^-(Y \setminus V) = X$ or $F^-(Y \setminus V) = \emptyset$. The result follows.
3. Let F^- be image-quasi-open and V be open. Then $F^-(Y \setminus V) \in \mathcal{QC}(X)$ because F is u-q-c and $F^-(Y \setminus V) = \cup_{y \in Y \setminus V} F^-(y) \in \mathcal{QO}(X)$ because F^- is image-quasi-open. Then $F^-(Y \setminus V) = X$ or $F^-(Y \setminus V) = \emptyset$ and the result follows.
4. Let F be image-open and F^- be image-quasi-open. By 2.14(2), $F^-(Y \setminus V) \in \mathcal{QC}(X)$, also $F^-(Y \setminus V) \in \mathcal{QO}(X)$ since F^- is image-quasi-open. The result follows.

5.2 **Corollary**. Let X be quasi-connected and $F: X \to Y$ be quasi-continuous and onto Y such that $F(a)$ is quasi-connected in Y for some $a \in X$. Then Y is connected.

Proof. Let V be clopen in Y. Then V and $Y \setminus V$ are quasi-separated. By 4.4, either $F(a) \subset V$ or $F(a) \subset Y \setminus V$. By 5.1(1), either $F(X) \subset V$ or $F(X) \subset Y \setminus V$. Since F is an onto multifunction, it follows that $V = Y$ or $V = \emptyset$. The result follows from [12] Theorem 6.1.1.

5.3 Corollary. Let X be quasi-connected and $F: X \to Y$ be quasi-continuous and onto Y such that $F(a) \subset V$ for some $a \in X$, where V is a component of Y. Suppose that either F is image-open or V is open. Then Y is connected.

Proof. The set V is always closed. Suppose first that F is image-open. By 5.1(2), $F^{+}(V) = X$, that is, $F(X) = V$. Then $Y = V$, hence Y is connected. Suppose now that V is open. The result then follows from 5.1(1).

5.4 **Corollary**. Let $F: X \to Y$ be image-open and quasi-continuous such that either F is image-closed or F^{-} is image-quasi-open. Then F is constant on each quasi-component of X.

Proof. Assume that X is quasi-connected and that $a \in X$. Let $V = F(a)$. Suppose that F is image-closed. It follows from 5.1(1) that $F(X) \subset V$, thus $F(a) = F(X)$. Suppose now that F^{-} is image-quasi-open. It follows from 5.1(3) that $F(X) \subset V$, thus $F(a) = F(X)$. Let $\{X_\alpha | \alpha \in \Lambda\}$ be the quasi-components of X and let $F_\alpha = F|X_\alpha$. Since each F_α has the same properties as F, the result follows from the first part of this proof.

Consider 2.16 once again before reading the next result.

5.5 **Corollary**. Let X, Y and F be as in 5.4, and suppose that the quasi-components of X are quasi-open. Then F is graph-quasi-open.

Proof. Let $\{X_\alpha | \alpha \in \Lambda\}$ be the class of quasi-components of X and let $Y_\alpha = F(X_\alpha)$. Then

$$\begin{aligned} G(F) &= \{(x, y)\} | y \in F(x)\} \\ &= \cup_{\alpha\in\Lambda}\{(x, y) | x \in X_\alpha \text{ and } y \in Y_\alpha\} \\ &= \cup_{\alpha\in\Lambda}(X_\alpha \times Y_\alpha) \in \mathcal{QO}(X \times Y) \end{aligned}$$

from [20] Theorems 2 and 11.

5.6 **Corollary**. Let $F: X \to Y$ be image-open and quasi-continuous. Then the multifunction $T: X \to Y$, defined by $T(x) = \overline{F(x)}$ for every $x \in X$, is constant on each quasi-component of X.

Proof. Similar to that of 3.10.

5.7 **Corollary**. Let $F: X \to Y$ be u-q-c, F^{-} image-quasi-open and Y a T_1-space.

1. Then F is constant on each quasi-component of X.

2. Then the multifunction $T: X \to Y$, defined $T(x) = \underline{F(x)}$ for every $x \in X$, is constant on each quasi-component of X.

Proof. Assume that X is quasi-connected. By 2.13(2), F is l-q-c; thus F is quasi-continuous and so 5.1 may be applied. Let $a \in X$.

1. Similar to that of 3.12(1)

2. Let $\mathcal{U} = \{U \in \mathcal{O}(Y) | T(a) \subset U\}$. By 5.1(3), and since $F(a) \subset U$, we deduce that $F(X) \subset \cap\mathcal{U}$. Since Y is T_1 it follows that $\cap\mathcal{U} \subset T(a)$, thus $\underline{F(X)} \subset T(a)$. Since (δ)

$$\underline{F(X)} = \underline{\cup_{x\in X}F(x)} \supset \cup_{x\in X}\underline{F(x)} = \cup_{x\in X}T(x) = T(X) \supset T(a),$$

it follows that $T(a) = \underline{F(X)}$ for any $a \in X$.

The example below shows that some of the hypotheses in the preceeding results are indeed necessary.

5.8 **Example.**

1. Let X, $\mathcal{O}(X)$, Y, $\mathcal{O}(Y)$, F and V be as in 3.2 (see also 3.6). Then $\mathcal{O}(X) = \mathcal{QO}(X)$, $\mathcal{C}(X) = \mathcal{QC}(X)$, X is quasi-connected, F is image-open (hence image-quasi-open) and F^- is image-quasi-open. But 5.1 - 5.4 and 5.6 fail since F is not quasi-continuous.

2. Let $X = \{a, b, c, d\}$ and $\mathcal{O}(X) =$ $\{\emptyset, \{a\}, \{b\}, \{a, b\}, \{b, c\}, \{b, d\}, \{a, b, c\}, \{a, b, d\}, \{b, c, d\}, X\}$. Then $\mathcal{C}(X) =$ $\{\emptyset, \{a\}, \{c\}, \{d\}, \{a, c\}, \{a, d\}, \{c, d\}, \{a, c, d\}, b, c, d\}, X\}$, and $\mathcal{QO}(X) = \mathcal{O}(X)$, $\mathcal{QC}(X) = \mathcal{C}(X)$. Then X is not quasi-connected. Define $F : X \to X$ as follows: $F(a) = \{b, c\}$, $F(b) = F(c) = F(d) = \{a, b, d\}$. Then F is quasi-continuous on X, F is image-open, F^- is image-quasi-open and $V = \{b, c, d\}$ is clopen. Then 5.1, 5.2 and 5.3 all fail. For 5.2, take $F(a) = \{b, c\}$, which is quasi-connected. For 5.3, take $F(a) = \{b, c\} \subset \{b, c, d\}$, the latter set being a component (and a quasi-component). This example also illustrates the validity of 5.4 and 5.6, where F and T are both constant on the quasi-component $\{b, c, d\}$.

3. Let X, $\mathcal{O}(X)$, Y and $\mathcal{O}(Y)$ be as in 3.11(2) and let $F : X \to Y$ be as in 3.13(2). Then neither F nor T is constant on X.

6. IMAGE-NON-MINGLED MULTIFUNCTIONS

6.1 **Definition.** Let $F : X \to Y$ be a multifunction.

1. ([40]) F is said to be image-non-mingled if $F(x_1) \cap F(x_2) \neq \emptyset$ implies that $F(x_1) = F(x_2)$.

2. ([2]) F is said to be injective if $x_1 \neq x_2$ implies that $F(x_1) \cap F(x_2) = \emptyset$.

An injective multifunction is evidently image-non-mingled. See [13], and also [36], p. 228 for some properties of injective multifunctions.

Image-non-mingled multifunctions occur frequently in algebra and in analysis, since all equivalences, certain additive, and all linear relations [21], are image-non-mingled.

6.2 **Proposition ([30]).** A multifunction $F : X \to Y$ is image-non-mingled if and only if there exist an equivalence relation R on $F(X)$ and a function $f : X \to Y$ such that $F = R \circ f$.

Note that if $F : X \to Y$ is image-non-mingled, then the class $\{F(x) | x \in X\}$ forms a partition of $F(X)$ which determines an equivalence relation R on $F(X)$ such that $R[y] = F(x)$ if $y \in F(x)$. By the Axiom of Choice, there exists a selector $f : X \to Y$ for F. Thus, $F(x) = R(f(x)) = (R \circ f)(x)$ for every $x \in X$.

6.3 **Proposition.** Let $F : X \to Y$ be a multifunction.

1. ([40]) F is image-non-mingled if and only if $F \circ F^- \circ F = F$.

2. ([30]) F is image-non-mingled if and only if its weak inverse $F^- : Y \to X$ is image-non-mingled.

6.4 **Theorem**. Let $F : X \to Y$ be image-non-mingled such that either F is image-open and l-q-c or F^- is image-quasi-open. Then F is u-q-c.

Proof. Let $a \in X, O \in \mathcal{O}(Y)$ such that $F(a) \subset O$ and let $U \in \mathcal{U}(a)$. Suppose first that F is image-open and l-q-c. We must show that F is u-q-c. Now, $F^-(F(a)) \in \mathcal{U}_q(a)$. If $V = U \cap F^-(F(a))$, then $V \in \mathcal{U}_q(a)$, by 2.7(1). There exists a non-empty set $W \in \mathcal{O}(X)$ such that $W \subset V$, see 2.1(1). Then, by 6.3(1),

$$F(W) \subset F(V) = F(U \cap F^-(F(a))) \subset F(F^-(F(a))) = F(a).$$

Thus, F is u-q-c. Suppose now that F^- is image-quasi-open. Then $F^-(F(a)) \in \mathcal{U}_q(a)$. By 2.13(2), we have that F is l-q-c. Proceed as above.

6.5 **Corollary**. Let $F : X \to Y$ be image-quasi-open and image-non-mingled.

1. Then F^- is quasi-continuous.

2. If either F^- is image-open or X is a T_1-space, then F^- is constant on each quasi-component of Y.

Proof.

1. By 6.3(2), F^- is image-non-mingled. Also, $F = (F^-)^-$ is image-quasi-open. Apply now 2.13(2) and 6.4 to F^-.

2. By (1) above, F^- is quasi-continuous. Suppose that F^- is image-open on X. Apply 5.4 to $F^- : Y \to X$ and recall that $F = (F^-)^-$. The result follows. Suppose now that X is a T_1-space. Apply 5.7(1) to F^-.

6.6 **Theorem**. Let $F : X \to Y$ be image-non-mingled such that either F is image-open and F^- is image-quasi-open, or F^- is image-quasi-open and Y a T_1-space. Then F is constant on each quasi-component of X.

Proof. First suppose that F is image-open and F^- is image-quasi-open. By 2.13(2) and 6.4, F is quasi-continuous. The result then follows from 5.4. Suppose now that F^- is image-quasi-open and that Y is a T_1-space. Then F is u-q-c by 6.4 and the result follows from 5.7(1).

6.7 **Theorem**. Let $F : X \to Y$ be image-non-mingled, F^- image-quasi-open and Y a T_1-space. Then the multifunction $T : X \to Y$, defined by $T(x) = \underline{F(x)}$ for every $x \in X$, is constant on each quasi-component of X.

Proof. This follows from 6.4 and 5.7(2).

We now derive a result on selectors for certain image-open multifunctions.

6.8 **Proposition**. Let $F: X \to Y$ be image-open, image-non-mingled, l-q-c and onto Y. Then F has a quasi-continuous selector $f: X \to Y$.

Proof. The class $\{F(x)|x \in X\}$ is a partition of Y. Denote this partition by the disjoint class $\{A_\alpha|\alpha \in \Lambda\}$, where $Y = \cup_{\alpha\in\Lambda} A_\alpha$. Assume the Axiom of Choice and select an element $y_\alpha \in A_\alpha$ for every $\alpha \in \Lambda$. Let $B = \{y_\alpha|\alpha \in \Lambda\}$. Let $[x_\alpha] = F^-(A_\alpha)$ for every $\alpha \in \Lambda$. Then $[x_\alpha] \in \mathcal{QO}(X)$, $[x_\alpha] = \{x \in X|F(x) = A_\alpha\}$ and $[x_{\alpha_1}] \cap [x_{\alpha_2}] = \emptyset$ if $\alpha_1 \neq \alpha_2$. Define $f: X \to Y$ by $f(x) = y_\alpha$ for every $x \in [x_\alpha]$. Let $V \in \mathcal{O}(Y)$ and let $\Lambda_1 = \{\alpha \in \Lambda|y_\alpha \in V\}$. Then $f^{-1}(V) = \cup_{\alpha\in\Lambda_1}[x_\alpha] \in \mathcal{QO}(X)$. Hence, f is quasi-continuous by 2.9.

6.9 **Remark**. If $F: X \to Y$ is graph-open, image-non-mingled and onto Y, then F has a continuous selector $f: X \to Y$. This follows from 2.15 and from 6.8.

In the following variation of 2.18, we relax the condition on Y but replace the pre-quasi-openness of F by a more stringent condition in order to obtain the lower-quasi-continuity of F. Compare this result with 6.4.

6.10 **Theorem**. Let $F: X \to Y$ be image-open, image-non-mingled and u-q-c. Then F is l-q-c.

Proof. Let $a \in X, O \in \mathcal{O}(Y)$ such that $F(a) \cap O \neq \emptyset$ and let $U \in \mathcal{U}(a)$. Now, $F^+(F(a)) \in \mathcal{U}_q(a)$ and $F^+(F(a)) = \{x \in X|F(x) = F(a)\}$. Let $V = U \cap F^+(F(a))$. Then $V \in \mathcal{U}_q(a)$ by 2.7(1). Let $W \in \mathcal{O}(X)$, $W \neq \emptyset$, such that $W \subset V$. Clearly, $W \subset F^-(O)$ and the result follows.

The next example shows that the condition that F be injective in [13] Theorem 7(2) cannot be replaced by the less stringent condition that F be image-non-mingled.

6.11 **Example**. Equip $I\!R$ with its usual topology and $I\!R^2$ with the discrete topology. Define $F: I\!R \to I\!R^2$ by

$$\begin{aligned} F(x) &= \{(0, 0)\} \text{ if } x \geq 0 \\ &= \{(\tfrac{1}{|x|}, -1)\} \text{ if } x < 0. \end{aligned}$$

Then Y is regular, F is pre-quasi-open, quasi-continuous (as required in [13] Theorem 7(2)), but not injective. However, F is image-non-mingled, but that is not sufficient to ensure the lower-semi-continuity of F on $I\!R$.

The last result of this paper shows that a sufficient condition for a l-q-c multifunction to be l-s-c is that the domain space is Baire.

6.12 **Theorem**. Let Y be a second countable space, X be a Baire space and F: $X \to Y$ be image-open, image-non-mingled and l-q-c. Then F is l-s-c.

Proof. Let $A = \{x \in X|F$ is not l-s-c at $x\}$. By[13] Theorem 16, A is of the first category in X. Suppose $A \neq \emptyset$ and let $a \in A$. Let $B = \{x \in X|F(x) = F(a)\}$. Then $B = F^-(F(a)) \in \mathcal{U}_q(a)$. Since $B \neq \emptyset$, it follows from 2.1(1) that $B^\circ \neq \emptyset$. It is easily seen that $B \subset A$. But then $A^\circ \neq \emptyset$, which is impossible since X is a Baire space. Thus, $A = \emptyset$, so that F is l-s-c on X.

REFERENCES

1. J.C. Bellenger and S. Simons, ***On the zeros of strict multifunctions,*** Preprint, 22 July 1986.

2. C. Berge, ***Topological spaces***, Oliver and Boyd, Edinburgh-London, 1st English edition, 1963.

3. N. Biswas, ***On some mappings in topological spaces***, Bull. Calcutta Math. Soc. 61(1969), 127-135.

4. S.E. Bohn and Lee Jong, ***Semi-topological groups***, Amer. Math. Monthly 72(1965), 996-998.

5. J. Ceder and S. Levi, ***On the search for Borel 1 selections***, Časopis Pěst. Mat. 110(1985), 19 - 32.

6. G. Choquet, ***Convergences***, Ann. Univ. Grenoble Sect. Sci. Math. Phys. (N.S.) 23(1948), 57 - 112.

7. S.G. Crossley and S.K. Hildebrand, ***Semi-closure***, Tex. J. Sci. 22(1971), 99 - 112.

8. —, ***Semi-closed sets and semi-continuity in topological spaces***, Tex. J. Sci. 22(1971), 123 - 126.

9. —, ***Semi-topological properties***, Fund. Math. 74(1972), 233 - 254.

10. P. Das, ***Note on semi-connectedness***, Indian J. Mech. Math. 12(1974), 31 - 34.

11. C. Dorsett, ***Semi-connectedness***, Indian J. Mech. Math. 17(1979), 57 - 63.

12. R. Engelking, ***General Topology***, Polish Scientific Publishers, Warsaw, 1977.

13. J. Ewart and T. Lipski, ***Quasi-continuous multivalued mappings***, Math. Slovaca 33(1983), 69 - 74.

14. S.P. Franklin, ***Open and image-open relations***, Colloq. Math. 12(1964), 209 - 211.

15. D. Gale and A. Mas-Colell, ***An equilibrium existence theorem for a general model without ordered preferences***, J. Math Econom. 2(1975), 9 - 15. Correction, J. Math. Econom. 6(1979), 297 - 298.

16. R. Henstock, ***Theory of Integration***, Butterworth, London, 1963

17. —, *A Riemann-type integral of Lebesgue power*, Canadian Journal of Math., 20(1968), 79 - 87.

18. S.Kempisty, *Sur les fonctions quasicontinues*, Fund. Math. 19(1932), 184 - 197.

19. J Kurzweil, *Generalized ordinary differential equations and continuous dependence on a parameter*, Czechoslovak Math. Jour. 7(82) (1957), 418 - 446.

20. N. Levine, *Semi-open sets and semi-continuity in topological spaces*, Amer. Math. Monthly 70(1963), 36 - 41.

21. S MacLane, *An algebra of additive relations*, Proc. Nat. Acad. Sci. U.S.A. 47(1961), 1043 - 1051.

22. S.N. Maheshwari and R. Prasad, *Some new separation axioms*, Ann. Soc. Sci. Bruxelles Sér. I 89(1975), 395 - 402.

23. P. Maritz, *Almost Baire class one multi-functions*, Bull. Austral. Math. Soc. 34(1986), 297 - 308.

24. N.F.G. Martin, *Quasi-continuous functions on product spaces*, Duke Math. J. 28(1961), 39 - 44.

25. A. Mas-Colell, *A selection theorem for open graph correspondences with star-shaped values*, J. Math. Anal. Appl. 68(1979), 273 - 275.

26. J.F. McClendon, *Subopen multifunctions and selections*, Fund. Math. 121 (1984), 25 - 30.

27. —, *On non-contractible valued multifunctions*, Pacific J. Math. 115(1984), 155 - 163.

28. R.M. McLeod, *The generalized Riemann integral*, The Carus Mathematical Monographs 20, MAA, 1980.

29. E.L. McShane, *Unified Integration*, Academic Press, New York, 1983.

30. A. Münnich and A. Száz, *An alternative theorem for continuous relations and its applications*, Publ. Inst. Math. (Beograd) (N.S.) 33(47) (1983), 163 - 168.

31. T. Neubrunn, *On Blumberg sets for multifunctions*, Bull. Acad. Polon. Sci. Sér. Sci. Math. 30(1982), 109 - 113.

32. —, *On quasicontinuity of multifunctions*, Math. Slovaca 32(1982), 147 - 154.

33. W.F. Pfeffer, ***The Riemann-Stieltjes approach to integration***, TWISK 187,CSIR, 1980.

34. V.I. Ponomarev, ***A new space of closed sets and many-valued continuous mappings of bicompacts***, Mat. Sb. (N.S.) 48(90) (1959), 191 - 212.

35. V. Popa, ***On a decomposition of the quasicontinuity of multifunctions***, Stud. Cerc. Mat. 27(1975), 323 - 328.

36. B. Ricceri, ***On multifunctions of one real variable***, J. Math. Anal. Appl. 124(1987), 225 - 236.

37. W.J. Shafer and H.F. Sonnenschein, ***Equilibrium in abstract economies without ordered preferences***, J. Math. Econom. 2(1975), 345 - 348.

38. M.S. Stanojević, ***On multivalued quotient mappings***, Publ. Inst Math. (Beograd) (N.S.) 17(31(1974), 155 - 161.

39. L.A. Steen and J.A. Seebach (Jr.), ***Counterexamples in topology***, Springer, New York, 2nd edition, 1978.

40. G.T. Whyburn, ***Continuity of multifunctions***, Proc. Nat. Acad. Sci. U.S.A. 54(1965), 1494 - 1501.

41. Yusufov, V.Sh., ***Open mappings***, Uspekhi Mat. Nauk 41 (1986), 185 - 186.

University of Stellenbosch,
Stellenbosch,
South Africa.

Infinite-dimensional generalised Riemann integrals

P.Muldowney

Abstract. *We show how to define a gauge, and hence an integral, in function space. Applying the theory to Feynman integration, we show that the Feynman integrals are path integrals.*

In defining the generalised Riemann integral in one dimension, we use intervals I in $\mathbb{R}$ of the form

$$]-\infty, a[, \ [u, v[, \ [b, \infty[$$

and define $|I|$ as

$$a, \ v-u, \ b$$

respectively. I and x are associated in $\mathbb{R}$, forming an associated interval-point pair (I, x), if x is

$$-\infty, \ u \text{ or } v, \ +\infty$$

respectively. A gauge is a positive function $\delta(x)$ defined for every $x \in \mathbb{R}^* = \mathbb{R} \cup \{-\infty, \infty\}$. Given an associated pair (I, x), then (I, x), or simply I, is δ-fine if $|I|$

$$< -1/\delta(x), \ < \delta(x), \ > 1/\delta(x)$$

respectively.

In n dimensions, an interval I is $I_1 \times \ldots \times I_n$, where each I_j is an interval of $\mathbb{R}$. $|I|$ is the product over j of $|I_j|$. If $x = (x_1, \ldots, x_n)$, $x_j \in \mathbb{R}^*$, then (I, x) is an associated pair if I_j, x_j are associated in $\mathbb{R}$, $1 \leq j \leq n$. Given $\delta(x) > 0$ and an associated pair (I, x), then (I, x), or, simply, I, is δ-fine if (I_j, x_j) is δ-fine, $1 \leq j \leq n$. (That is, $|I_j| < -1/\delta(x)$, $< \delta(x)$, $> 1/\delta(x)$ for $x_j = -\infty$, $x_j \in \mathbb{R}$, $x_j = +\infty$, respectively.

A partition of $\mathbb{R}^n$ is a finite collection of disjoint intervals I whose union is $\mathbb{R}^n$. A division of $\mathbb{R}^n$ is a finite collection of associated (I, x) such that the intervals I form a partition of $\mathbb{R}^n$. A division is δ-fine if each (I, x) in the division is δ-fine.

Given a point function $h(x)$ and a division $\mathcal{E} = \{(I, x)\}$, the Riemann sum is denoted by

$$(\mathcal{E}) \sum h(x)|I|$$

or simply $\sum h(x)|I|$, where we take $h(x)$ equal to zero by definition if x has any infinite components. Similarly, if $h(I,x)$ is a function of associated (I,x), then the Riemann sum is $\sum h(I,x)$, making a similar provision whenever x is a point at infinity.

For the infinite - dimensional case, let $]\tau',\tau[$ be an interval of $\mathbb{R}$, and let $T = \mathbb{R}^{]\tau',\tau[}$. Thus T is the set

$$\{x : t \mapsto x(t),\ t \in]\tau',\tau[,\ x(t) \in \mathbb{R}\}.$$

Let

$$\begin{aligned}
N &= \{t_1,\ldots,t_n\},\ t_j \in]\tau',\tau[,\ 1 \le j \le n,\\
x_j &= x(t_j),\ 1 \le j \le n,\\
x(N) &= (x_1,\ldots,x_n),\\
I &= I[N] = \{x : x(t_j) \in I_j,\ 1 \le j \le n\} \text{ each } I_j \text{ being an interval of } \mathbb{R},\\
I(N) &= I_1 \times \ldots \times I_n,\\
|I| &= |I[N]| = \textstyle\prod_{j=1}^n |I_j|.
\end{aligned}$$

Thus $I[N]$ is infinite-dimensional and is a subset of T, whereas $I(N)$ is finite-dimensional and is the projection of $I[N]$ into a space $\mathbb{R}^n$. Given $I[N]$, then $I_j = I(t_j)$ will denote the projection of I into dimension t_j. $I[N]$, x are associated in T if $I(N)$, $x(N)$ are associated in $\mathbb{R}^n$. Thus the triple I, N, x are associated if N is the dimension set in which I is restricted, and, for $t_j \in N$, $x(t_j)$ is associated with $I(t_j)$. The definitions of partition of T and division of T are analogous to the corresponding definitions of partition and division of $\mathbb{R}^n$.

A gauge in T is a device which restricts or reduces the volume $|I| = |I[N]|$ by, ***firstly***, increasing the dimension set N in which I is restricted, and, ***secondly***, reducing the edges $I(t)$, $t \in N$.

We illustrate this by means of diagrams.

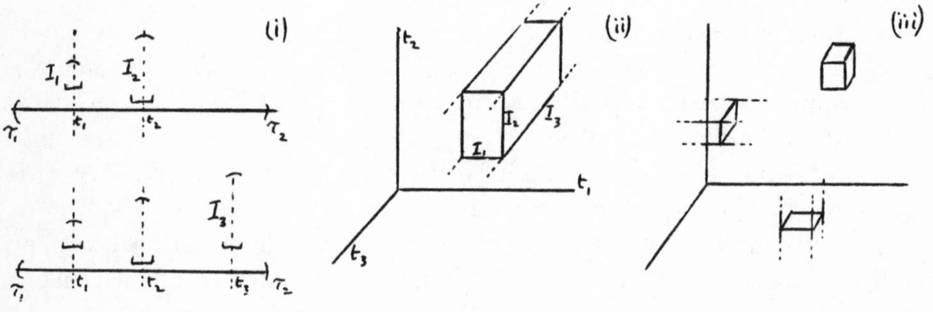

Diagram (iii) shows that the space can be filled by non-intersecting intervals, some of which are restricted in a particular dimension while others are unrestricted in that dimension. Thus the intervals of a division may have different dimension sets. (A remark to the contrary appears in [2] p.21.)

We consider three ways of defining a gauge in function space:

(a) A gauge can be defined by means of a δ-function, just as in finite-dimensional spaces. Let $\delta(x) > 0$ for $x \in T^*$. Let γ_a denote $\{\delta(x) : x \in T^*\}$. If I, N, x are associated then $(I[N], x)$, or $I[N]$, is γ_a-fine if $\max(t_j - t_j) < \delta(x)$ for $t_j, t_{j-1} \in N$, and $I(N)$ is δ-fine.

(b) Let A be a countable subset of $]\tau', \tau[$. For each $x \in T^*$ let $L(x)$ be a finite subset of A, and for each finite $N = \{t_1, \ldots, t_n\}$ let $\delta(y)$ be a gauge in $\mathbb{R}^n$. Let $\gamma_b = \{(A, L(x), \delta(x(N))) : x \in T^*\}$. If I, N, x are associated then $(I[N], x)$, or $I[N]$, is γ_b-fine if $N \supseteq L(x)$ and $I(N)$ is $\delta(x(N))$-fine.

(c) Omit the countable set A from the previous definition, so that, for each x, $L(x)$ is an arbitrary finite dimension-set. A gauge γ_c is

$$\{(L(x), \delta(x(N))) : x \in T^*\}.$$

Proposition 1. There exists a γ_a-fine division of T.

Proof. Without loss of generality we can take $\tau - \tau' \leq 1$. Let p be a positive integer and let $j \in \{1, 2, \ldots, 2^p - 1\}$. Let $\tau_j^{(p)} = \tau' + (\tau - \tau')j2^{-p}$. Let $q \in \{-p2^p, -p2^p + 1, \ldots, p2^p - 1\}$. Let $\mathcal{J}$ denote the family of intervals of the form

$$]-\infty, -p[, \ [q2^{-p}, (q+1)2^{-p}[, \ [p, \infty[.$$

Let

$$I^{(p)} = \{x : x(\tau_j^{(p)}) \in J \in \mathcal{J}, j = 1, \ldots, 2^p - 1\}.$$

Then I^{p+1} is a subset of I^p or $I^{p+1} \cap I^p = \emptyset$. Suppose there is no γ_a-fine division of T. Then, given p, there exist I^p and I^{p+1} for which there are no γ_a-fine divisions. Using the Cartesian product topology in T, the sequence $\ldots \mathrm{Cl}I^p \supset \mathrm{Cl}I^{p+1} \ldots$ has the finite intersection property so there exists $x \in \bigcap_{p=1}^{\infty} \mathrm{Cl}I^p$. Now choose p so that $2^{-p} < \delta(x)$ and $p > 1/\delta(x)$. Thus the assumption that no γ_a-fine division exists gives a contradiction.

In general, proofs of the existence of γ-fine divisions follow this pattern of bisection.

Proposition 2. There exists a γ_b-fine division of T.

Proof. See [1], and [2] Prop.1 p.20.

R. Henstock proves in these Proceedings that γ_c-fine divisions exist.

Definition. Let h be a real or complex-valued functional of associated triples I, x, N. Then $h(I, x, N)$ is a-integrable with $\int_T h(I, x, N) = \omega$ if, given $\varepsilon > 0$, there exists a gauge γ_a so that, for every γ_a-fine division of T,

$$\sum |h(I, x, N) - \omega| < \varepsilon.$$

Similarly, we define b- and c-integrability relative to gauges γ_b, γ_c. Let $\mathcal{H}_a, \mathcal{H}_b, \mathcal{H}_c$ denote the classes of functionals which are integrable relative to gauges $\gamma_a, \gamma_b, \gamma_c$, respectively.

Proposition 3. $\mathcal{H}_a \subseteq \mathcal{H}_b \subseteq \mathcal{H}_c$.

Proof. Every gauge of the type γ_b is also a gauge of type γ_c, so $\mathcal{H}_b \subseteq \mathcal{H}_c$. Given $\gamma_a = \{\delta(x) : x \in T^*\}$, choose A and $L(x) = \{\tau_1, \ldots, \tau_m\}$ so that $\tau_j - \tau_{j-1} < \delta(x)$, $j = 2, 3, \ldots, m$, and let $\delta(x(N)) < \delta(x)$. Thus, for each γ_a, there exists γ_b so that if a division is γ_a-fine then $\mathcal{E}$ is γ_b-fine. Therefore $\mathcal{H}_a \subseteq \mathcal{H}_b$.

Let $\lambda = \mu + \iota\nu$ be a complex number, where ι denotes the square root of -1, let

$$w_\lambda(x, N) = \exp(\lambda \sum_{j=1}^{n} \frac{(x_j - x_{j-1})^2}{t_j - t_{j-1}} \prod_{k=1}^{n} (\frac{-\pi(t_j - t_{j-1})}{\lambda})^{-1/2},$$

and let $w_\lambda(I, x, N) = w_\lambda(x, N)|I[N]|$. This is integrable (relative to all three types of gauge, $\gamma_a, \gamma_b, \gamma_c$) provided $\mu \leq 0$, $\nu \geq 0$, $\lambda \neq 0$. See [2] Prop. 68 p.84, where the proof of integrability refers to gauges γ_b, but can easily be adapted to gauges γ_a, γ_c. From here on, every gauge γ will be a gauge of type γ_b. If $\nu = 0$, $\mu < 0$ we call $w_\lambda = w_\mu$ the Wiener integrator, and if $\mu = 0$, $\nu > 0$, then $w_\lambda = w_{\iota\nu}$ is the Feynman integrator.

Let α, θ be real numbers with $\alpha < 1/2$, $\theta > 0$ and let

$$\begin{aligned} C(\alpha, \theta) &= \{x : |x(t) - x(t')|^\alpha < \theta|t - t'| \text{ as } t \to t',\ t, t' \in [\tau', \tau]\}, \\ D(\alpha, \theta) &= T \setminus C(\alpha, \theta). \end{aligned}$$

It is proven in [2] p.103 that, for $\mu < 0$, $\nu \geq 0$, we have

$$\int_T \chi(D(\alpha, \theta), x) w_{\mu+\iota\nu}(I, x, N) = 0,$$

where χ denotes the indicator functional for subsets of T. This shows that those $x \in T$ which are not continuous functions, or paths, contribute nothing to the integral. Thus $\int_T w_\lambda(I, x, N) = \int_C w_\lambda(I, x, n)$ is an integral over the space C of paths, or path integral. We wish to let $\mu \to 0-$ and establish that this also holds for $\mu = 0$, $\nu > 0$, i.e. that the Feynman integrals are path integrals. The key to this is that, for fixed $\nu > 0$ and given x, N, the functional $|w_\lambda(x, N)|$ is monotone increasing as $\mu \to 0-$, except for the case $x_1 = x_2 = \ldots = x_n$, when this expression is monotone decreasing. We can avoid this exceptional case by using a different but equivalent form of the generalised Wiener integrator, namely,

$$w_\lambda(I, N) = \int_{I(N)} \exp \sum_{j=1}^{n} \frac{\lambda(y_j - y_{j-1})^2}{t_j - t_{j-1}} dy_1 \ldots dy_{n-1} \prod_{j=1}^{n} (\frac{-\pi(t_j - t_{j-1})}{\lambda})^{-1/2}.$$

This expression is variationally equivalent to $w_\lambda(I,x,N)$, [2] p.84. As the set of $y \in I(N)$ satisfying $y_1 = \ldots = y_n$ is null relative to generalised Riemann integration in $I(N)$, we have $|w_\lambda(I,N)|$ monotone increasing as $\mu \to 0-$.

Proposition 4. If $\nu > 0$, $\alpha < 1/2$, $\theta > 0$ then

$$\int_T \chi(D(\alpha,\theta),x)|w_{\iota\nu}(I,N)| = 0.$$

Proof. $w_{\mu+\iota\nu}(I,N) \to w_{\iota\nu}(I,N)$ as $\mu \to 0-$. For each I, N there exists μ_0 so that $\mu_0 < \mu' < \mu < 0$ implies

$$|w_{\mu'+\iota\nu}(I,N)| < |w_{\mu+\iota\nu}(I,N)|.$$

As $w_{\iota\nu}$ is of generalised bounded variation ([2] Prop. 85 p.98), there exist $\beta_j > 0$ and subsets X_j of T whose union is T such that

$$\sum |w_{\iota\nu}(I,N)|\chi(X_j,x) < \beta_j$$

for all γ-fine divisions, $j = 1, 2, \ldots$. If $x \in X_j$, choose μ_1 so that, if $\mu_1 < \mu < 0$ and $(I[N],x)$ is γ-fine, then

$$|w_{\iota\nu}(I,N)| - |w_{\mu+\iota\nu}(I,N)| < \varepsilon |w_{\iota\nu}(I,N)|\beta_j^{-1}2^{-j}.$$

Take $g_0(I,x,N) = \sum_{j=1}^{\infty} |w_{\iota\nu}(I,N)\beta_j^{-1}2^{-j}\chi(X_j,x)$. g_0 is then a functional of bounded variation. The result now follows from Levi's monotone convergence theorem [2] Prop.26 p.39. (In the statement of Prop.26 we have g_0 integrable. However, the proofs of this and consequent results do not actually require the integrability of g_0. It is clearly sufficient that g_0 be of bounded variation, or simply have bounded Riemann sums.)

References

1. Henstock, R., 'Integration in product spaces, including Wiener and Feynman integration', Proc. London Math. Soc. 27 (1973) 317-344.

2. Muldowney, P., A General Theory of Integration in Function Spaces, Longman, 1987.

Magee College
University of Ulster
N.Ireland

The space of Henstock integrable functions II

Piotr Mikusiński, and Krzysztof Ostaszewski

Abstract

The space of Henstock integrable functions on the unit cube in the m-dimensional Euclidean space is normed, barrelled, and not complete. We describe its completion in the space of Schwartz distributions.

We also show how the distribution functions for finite signed Borel measures are multipliers for the Henstock integrable functions, and how they generate continuous linear functionals on the space of Henstock integrable functions. Finally, we discuss various integration by parts formulas for the two-dimensional Henstock integral.

1980 Mathematics Subject Classification
Primary 46E10 Secondary 26A39

Key words and phrases
Henstock integral, distribution, completion, continuous linear functional, integration by parts.

1.1. Definition. Let $I_0 \subset \mathbb{R}^m$ be the unit cube in the m-dimensional Euclidean space. A function $f : I_0 \to \mathbb{R}$ will be termed ***Henstock integrable***, with

$$\int\int \cdots \int_{I_0} f(x_1, x_2, \ldots, x_m) dx_1 dx_2 \ldots dx_m \tag{1}$$

written for the value of the integral, if for every $\varepsilon > 0$ there exists a positive function $\delta : I_0 \to \mathbb{R}$ (usually called a ***gauge***) such that whenever

$$\pi = \{((x_1^i, x_2^i, \ldots, x_m^i), I_i) : i = 1, 2, \ldots, n\} \tag{2}$$

is a partition of I_0, consisting of pairs of points in I_0 and nonoverlapping subintervals of I_0 whose union is the whole I_0, and such that for every $i = 1, 2, \ldots, n$, $(x_1^i, x_2^i, \ldots, x_m^i) \in I_i$ and I_i is contained in the ball centered at $(x_1^i, x_2^i, \ldots, x_m^i)$ of radius $\delta(x_1^i, x_2^i, \ldots, x_m^i)$, we have

$$|\sum_{i=1}^{n} f(x_1^i, x_2^i, \dots, x_m^i)\lambda(I_i) - \int\int\dots\int_{I_0} f(x_1, x_2, \dots, x_m)dx_1dx_2\dots dx_m| < \varepsilon; \tag{3}$$

here λ stands for the m-dimensional volume of an interval $I \subset I\!R^m$. Quite often we will simply write

$$\int_{I_0} fd\lambda \tag{4}$$

for the Henstock integral of f over I_0.

A partition π as in (2), satisfying conditions listed between (2) and (3) will be called ***δ-fine***.

We will denote by $\tilde{f}$ the indefinite Henstock integral of a function f, i.e.,

$$\tilde{f}(x_1, x_2, \dots, x_m) = \int_0^{x_1}\int_0^{x_2}\dots\int_0^{x_m} f(t_1, t_2, \dots, t_m)dt_1dt_2\dots dt_m =$$

$$\int\int\dots\int_{I_0} f(t_1, t_2, \dots, t_m)\chi_{[0,\,x_1]\times[0,\,x_2]\times[0,\,x_m]}dt_1dt_2\dots dt_m, \tag{5}$$

where χ_E denotes the characteristic function of set $E \subset I\!R$.

If H is an interval function and we replace λ by H in (3) then we get the concept of the Henstock integral of f with respect to H, written as

$$\int\int\dots\int_{I_0} fdH \text{ or just } \int_{I_0} fdH. \tag{6}$$

If $g:\ I_0 \to I\!R$ then it generates an interval function H as follows. Let

$$I = [a_1, b_1] \times [a_2, b_2] \times \dots \times [a_m, b_m], \tag{7}$$

define

$$H(I) = \sum_{J\subset\{1,2,3,\dots,m\}} (-1)^{\mathrm{card}J} g(c_1, c_2, \dots, c_m), \tag{8}$$

where

$$\{c_1, c_2, \dots, c_m\} = \{a_{j_1}, a_{j_2}, \dots a_{j_k}, b_{i_1}, b_{i_2}, \dots, b_{i_l}\}, \tag{9}$$

and

$$\{j_1, j_2, \dots, j_k\} = J,\ \{i_1, i_2, \dots, i_l\} = \{1, 2, \dots, m\} \setminus J, \tag{10}$$

and card$J$ is the cardinality of J. The integral

$$\int\int\dots\int_{I_0} fdH \tag{11}$$

is called the Henstock integral of f with respect to g and written as

$$\int\int\cdots\int_{I_0} f dg \text{ or simply } \int_{I_0} f dg. \tag{12}$$

Please consult [2], [4], [5], and [7] for these definitions.

1.2. The class of Henstock integrable functions on I_0 will be denoted by $\mathcal{H}$. It is a linear topological space. In [8] and [9] it is shown that the space equipped with the Alexiewicz norm is barrelled, but it is not a Banach space. [6] and [8] discuss the dual of the space. The work in [8] is done in the two-dimensional case, but easily extends to the multidimensional one. [6] considers the dual of $\mathcal{H}$ for functions of one variable.

Our intention is to describe the completion of the space and to further discuss its dual.

Let us note that every Henstock integrable function $f : I_0 \to \mathbb{R}$ is a Schwartz distribution (see [4], section 2.12).

1.3. Definition. Denote by $\mathcal{F}$ the space of all ***distributions of order*** m with support in I_0, i.e., $f \in \mathcal{F}$ if there exists a continuous function $F : \mathbb{R}^m \to \mathbb{R}$ such that

$$F(x_1, x_2, \ldots, x_m) = 0 \text{ if } \min\{x_1, x_2, \ldots, x_m\} \le 0, \tag{13}$$

$$F(x_1, x_2, \ldots, x_i, \ldots, x_m) = F(x_1, x_2, \ldots, 1, \ldots, x_m) \text{ if } x_i \ge 1 \tag{14}$$

for $i = 1, 2 \ldots, m$,

$$f = \frac{\partial^m F}{\partial x_1 \partial x_2 \ldots \partial x_m}, \tag{15}$$

where the derivatives are understood in the distributional sense.

For $f \in \mathcal{F}$ as in (13), (14), and (15) define

$$\int_0^{x_1}\int_0^{x_2}\cdots\int_0^{x_m} f(t_1, t_2, \ldots, t_m) dt_1 dt_2 \ldots dt_m = F(x_1, x_2, \ldots, x_m) \tag{16}$$

for $(x_1, x_2, \ldots, x_m) \in I_0$. Note that, for every $f \in \mathcal{F}$ there exists exactly one function F satisfying (13), (14), and (15). Thus the integral (16) is uniquely defined. Moreover

$$||f|| = \sup_{(x_1, x_2, \ldots, x_m)\in I_0} |F(x_1, x_2, \ldots, x_m)| \tag{17}$$

is a norm on $\mathcal{F}$. We will call it the ***Alexiewicz norm***, as it is the same as the Alexiewicz norm introduced in [8] on the space of Henstock integrable functions.

1.4. Proposition. *$\mathcal{F}$ is complete.*

Proof. Let $\{f_n\}$ be a Cauchy sequence in $\mathcal{F}$. Let

$$F_n(x_1, x_2, \ldots, x_m) = \int_0^{x_1} \int_0^{x_2} \cdots \int_0^{x_m} f_n(t_1, t_2, \ldots, t_m) dt_1 dt_2 \ldots dt_m. \quad (18)$$

Then $\{F_n\}$ is a Cauchy sequence (with respect to the sup norm) of continuous functions satisfying (13), (14), and (15), therefore it is convergent to a continuous function F satisfying the same conditions. Define

$$f = \frac{\partial^m F}{\partial x_1 \partial x_2 \ldots \partial x_m}. \quad (19)$$

Then $f \in \mathcal{F}$ and $\lim_{n\to\infty} ||f_n - f|| = 0$.

1.5. Observation. $\mathcal{H} \subset \mathcal{F}$.

Proof. This immediately follows from the following statement proved in [4], sections 2.3, and 2.12: if $f \in \mathcal{H}$ then $\tilde{f}$ is continuous and if $\phi : I_0 \to I\!R$ is m times continuously differentiable and

$$\psi(x_1, x_2, \ldots, x_m) = \frac{\partial^m \phi(x_1, x_2, \ldots, x_m)}{\partial x_1 \partial x_2 \ldots \partial x_m} \quad (20)$$

then

$$\int\int \cdots \int_{I_0} f(x_1, x_2, \ldots, x_m)\phi(x_1, x_2, \ldots, x_m) dx_1 dx_2 \ldots dx_m =$$

$$(-1)^m \int\int \cdots \int_{I_0} \tilde{f}(x_1, x_2, \ldots, x_m)\psi(x_1, x_2, \ldots, x_m) dx_1 dx_2 \ldots dx_m. \quad (21)$$

1.6.Theorem. *$\mathcal{F}$ is the completion of $\mathcal{H}$.*

Proof. Denote by $\mathcal{H}^*$ the completion of $\mathcal{H}$ with respect to the Alexiewicz norm. Then $\mathcal{H}^* \subset \mathcal{F}$ and both spaces are complete with respect to the same norm. Therefore, by the open mapping theorem, they are equal.

1.7. Remark. In the one-dimensional case it is known that every Henstock integrable function is almost everywhere a derivative of its indefinite integral. This implies that in that case, $\mathcal{H}$ is of the first category in $\mathcal{F}$. An easy example of an element of $\mathcal{F}$ which is not in $\mathcal{H}$ is in that case a distributional derivative of a nowhere differentiable continuous function.

2.1. We will turn now to our discussion of the dual of the space $\mathcal{H}$. We have the following, as presented in [6] and [8]:

In the one-dimensional case T is a continuous linear functional on $\mathcal{H}$ if and only if either of the following holds (all integrals used below are Henstock integrals):

(a) There exists a finite signed Borel measure μ_T on $(0, 1]$ such that

$$T(f) = \int_0^1 \tilde{f}(t) d\mu_T(t), \quad (22)$$

where, as usual

$$\tilde{f}(x) = \int_0^x f(t)dt. \tag{23}$$

(b) There exists a function $g_T : [0,\ 1] \rightarrow \mathbb{R}$ of essentially bounded variation such that

$$T(f) = \int_0^1 f(t)g_T(t)dt. \tag{24}$$

Being of essentially bounded variation is equivalent to having a signed finite Borel measure as a distributional derivative. If μ_g stands for that distributional derivative then integration by parts yields

$$\int_0^1 \tilde{f}(t)d\mu_T(t) = \int_0^1 \tilde{f}(t)d\mu_g(t) + \tilde{f}(1)\mu_g((0,\ 1]). \tag{25}$$

Notice that the expression $\tilde{f}(1)\mu_g((0,\ 1])$ is itself a continuous linear functional of f.

As observed in [8] the description (a) easily extends to the multidimensional case. However, (b) uses the class of multipliers for the Henstock integrable functions (i.e., functions which multiplied by a Henstock integrable function produce a Henstock integrable function), which is not known in the multidimensional case.

For simplicity, let us restrict ourselves to the two-dimensional case, with $I_0 = [0,\ 1] \times [0,\ 1]$. This does not affect generality of the results.

2.2. Definition A function $g :\ I_0 \rightarrow \mathbb{R}$ is of ***strongly bounded variation*** (see[4]) if for every $x \in [0,\ 1]$, $g(x,\ \cdot)$ is of bounded variation, for every $y \in [0,\ 1]$, $g(\cdot,\ y)$ is of bounded variation, and

$$\sup \sum_{i=1}^{n} |g(a_i,\ c_i) - g(a_i,\ d_i) - g(b_i,\ c_i) + g(b_i,\ d_i)| < +\infty, \tag{26}$$

where the least upper bound is taken over all partitions of I_0 into a finite collection of nonoverlapping nondegenerate closed intervals $[a_i,\ b_i] \times [c_i,\ d_i]$, $i = 1, 2, 3, \ldots, n$.

Let us note that [4] contains the definition of a function of strongly bounded variation in the general multidimensional case.

2.3. Theorem. ***Every function of strongly bounded variation is a multiplier for Henstock integrable functions.***

Proof. See[4].

2.4. It is not known whether the above is a complete characterization of multipliers. Our intention is to point out a specific subclass of the class of functions of strongly bounded variation.

2.5. Definition. Let $\mathcal{D}$ stand for the class of two-dimensional distribution functions of finite signed Borel measures on $(0\ ,1] \times (0,\ 1]$. For example, if μ is a positive measure then $g_\mu \in \mathcal{D}$ corresponding to it is

$$g_\mu(x, y) = \mu((0, x] \times (0, y]). \tag{27}$$

The value of $g_\mu(x, y)$ for $x = 0$ or $y = 0$ is inessential to us. We will assume it to be zero.

In general, for a signed finite Borel measure μ on $(0, 1] \times (0, 1]$ we will denote its distribution function by g_μ.

Also, we will denote by $\mathcal{M}$ the class of finite signed Borel measures on $(0, 1] \times (0, 1]$. $\mathcal{M}^+$ will denote the class of positive measures in $\mathcal{M}$.

2.6. Proposition. *The elements of $\mathcal{D}$ are of strongly bounded variation.*

Proof. It suffices to show that for a $\mu \in \mathcal{M}^+$, g_μ is of strongly bounded variation. One can easily see that both $g_\mu(x, \cdot)$ and $g_\mu(\cdot, y)$ are monotone for every $x \in [0, 1]$ and every $y \in [0, 1]$, so that they are of bounded variation. Let $\{I_i : i = 1, 2, \ldots, n\}$ be a finite class of nonoverlapping nondegenerate subintervals of I_0, and $I_i = [a_i, b_i] \times [c_i, d_i]$ for $i = 1, 2, \ldots, n$. Consider the sum

$$\sum_{i=1}^{n} |g_\mu(b_i, d_i) - g_\mu(a_i, c_i) - g_\mu(b_i, d_i) + g_\mu(a_i, b_i)| =$$

$$\sum_{i=1}^{n} |\mu((a_i, b_i] \times (c_i, d_i])| \leq ||\mu||. \tag{28}$$

This implies that g_μ is of strongly bounded variation.

2.7. Corollary. *A distribution function of a finite signed Borel measure is a multiplier for Henstock integrable functions.*

2.8. Corollary. *If $g : I_0 \to \mathbb{R}$ is equivalent to a distribution function of a finite signed Borel measure then g is a multiplier for Henstock integrable functions.*

2.9. Definition. Let $\mathcal{C}_0$ denote the class of all continuous $F : I_0 \to \mathbb{R}$ such that F is continuous and $F(x, y) = 0$ whenever $x = 0$ or $y = 0$.

2.10. Observation. *If $F \in \mathcal{C}_0$) and $\mu \in \mathcal{M}$ then the Lebesgue-Stieltjes integral*

$$\mathcal{L}\int\int_{I_0} F d\mu \tag{29}$$

exists and is well-defined.

2.11. Proposition. *The Riemann-Stieltjes integral of $F \in \mathcal{C}_0$ with respect to a $\mu \in \mathcal{M}$, denoted by*

$$\mathcal{R}\int\int_{I_0} F d\mu, \tag{30}$$

is naturally defined as the limit of the Riemann sums

$$\sum_{i=1}^{n} F(x_i, y_i)\mu((a_i, b_i] \times (c_i, d_i]), \tag{31}$$

where the elements of $\{(a_i, b_i] \times (c_i, d_i] : i = 1, 2, \ldots, n\}$ *are disjoint, their union is* $(0, 1] \times (0, 1]$, *and* $(x_i, y_i) \in (a_i, b_i] \times (c_i, d_i]$ *for* $i = 1, 2, \ldots, n$. *The limit is taken with respect to the norm of the partition, which is the diameter of the largest of the intervals in the partition, tending to zero.*

Existence of

$$(\mathcal{R}) \int\int_{I_0} F d\mu \tag{32}$$

implies existence of the Henstock integral of F *with respect to* g_μ *and their equality.*

Proof. It suffices to consider $\mu \in M^+$. Let $\varepsilon > 0$ be arbitrary. Choose a $\delta > 0$ such that whenever the norm of the partition

$$\{(a_i, b_i] \times (c_i, d_i] : i = 1, 2, \ldots,\} \tag{33}$$

is less than δ, and (x_i, y_i) belongs to $(a_i, b_i] \times (c_i, d_i]$ for every $i = 1, 2, \ldots, m$, we have

$$|\sum_{i=1}^{n} F(x_i, y_i)\mu((a_i, b_i] \times (c_i, d_i]) - (\mathcal{R}) \int\int_{I_0} F d\mu| < \varepsilon \tag{34}$$

Now let $p : I_0 \to \mathbb{R}$ be a gauge function defined as follows:

$$\begin{aligned} p(x, y) &= \delta/2 \text{ if } (x, y) \in (0, 1] \times (0, 1], \\ \text{and } &= 2 \text{ if } (x, y) \notin (0, 1] \times (0, 1]. \end{aligned} \tag{35}$$

If

$$\pi = \{((x_i, y_i), (a_i, b_i] \times (c_i, d_i]) : i = 1, 2, \ldots, n\} \tag{36}$$

is a p-fine partition, then its norm is less than δ and

$$\sum_{i=1}^{n} F(x_i, y_i)(g_\mu(a_i, c_i) - g_\mu(a_i, d_i) - g_\mu(b_i, c_i) + g\mu(b_i, d_i)) =$$

$$\sum_{i=1}^{n} F(x_i, y_i)\mu((a_i, b_i] \times (c_i, d_i]). \tag{37}$$

Consequently, the Henstock integral of F with respect to g_μ exists and equals

$$(\mathcal{R}) \int\int_{I_0} F d\mu \tag{38}$$

2.12. Proposition. *If* $F \in \mathcal{C}_0$ *and* $\mu \in \mathcal{M}$ *then*

$$(\mathcal{R}) \int\int_{I_0} F d\mu \tag{39}$$

exists.

Proof. Let $\varepsilon > 0$ be arbitrary. Choose a number η such that

$$\eta < \frac{\varepsilon}{2\mu((0,\ 1] \times (0,\ 1])}. \tag{40}$$

There exists a $\delta > 0$ such that if $(x_1,\ y_1), (x_2,\ y_2) \in I_0$ and the distance from $(x_1,\ y_1)$ to $(x_2,\ y_2)$ is less than δ then $|F(x_1,\ y_1) - F(x_2,\ y_2)| < \eta$. Let

$$\{((x_i,\ y_i),\ I_i) : i = 1,\ 2,\ \ldots,\ n\},\ \ \{((s_j,\ t_j),\ J_j) : j = 1,\ 2,\ \ldots,\ r\} \tag{41}$$

be two partitions of I_0, both with norm less than δ. Consider the intervals

$$P_{i,j} = I_i \cap J_j, i = 1,\ 2,\ \ldots,\ n,\ j = 1,\ 2,\ \ldots,\ r. \tag{42}$$

In each nonempty $P_{i,j}$ choose a point $(u_{i,j}, v_{i,j})$. Then the distances between $(u_{i,j}, v_{i,j})$ and $(x_i,\ y_i)$, and between $(u_{i,j},\ v_{i,j})$ and $(s_j,\ t_j)$ are both less than δ. Thus

$$|\sum_{i,j=1}^{n,r} F(u_{i,j},\ v_{i,j})\mu(P_{i,j}) - \sum_{i=1}^{n} F(x_i,\ y_i)\mu(I_i)| =$$

$$|\sum_{i=1}^{n}(\sum_{j=1}^{r} F(u_{i,j},\ v_{i,j})\mu(P_{i,j}) - F(x_i,\ y_i)\mu(P_{i,j}))| \leq \tag{43}$$

$$\sum_{i=1}^{n}\sum_{j=1}^{r} |F(u_{i,j},\ v_{i,j}) - F(x_i, y_i)|\mu(P_{i,j}) \leq \eta\mu(I_0) < \frac{1}{2}\varepsilon.$$

Similarly

$$|\sum_{i,j=1}^{n,r} F(u_{i,j},\ v_{i,j})\mu(P_{i,j}) - \sum_{j=1}^{r} F(s_j,\ t_j)\mu(J_j)| < \frac{1}{2}\varepsilon. \tag{44}$$

Therefore

$$|\sum_{i=1}^{n} F(x_i,\ y_i)\mu(I_i) - \sum_{j=1}^{r} F(s_j,\ t_j)\mu(J_j)| > \varepsilon, \tag{45}$$

which implies that

$$(\mathcal{R})\int\int_{I_0} F d\mu \tag{46}$$

exists.

2.13. Proposition. *If $F \in \mathcal{C}_0$ and $\mu \in \mathcal{M}$ then*

$$(\mathcal{L})\int\int_{I_0} F d\mu = (\mathcal{R})\int\int_{I_0} F d\mu. \tag{47}$$

Proof. Let $\{\pi_k\}_{k\in N}$ be a sequence of partitions of I_0,

$$\pi_k = \{((x_i^k,\, y_i^k),\, I_i^k) : i = 1,\, 2,\, \ldots,\, n_k\} \tag{48}$$

such that the norm of the partition π_k goes to zero as $k \to \infty$. For each $k \in \mathbb{N}$ define

$$U_k(x,\, y) = \sup_{(s,t)\in I_i^k} F(s,\, t) \text{ for } (x,\, y) \in I_i^k,\ i = 1,\, 2,\, \ldots,\, n_k,$$

$$L_k(x,\, y) = \inf_{(s,t)\in I_i^k} F(s,\, t) \text{ for } (x,\, y) \in I_i^k,\ i = 1,\, 2,\, \ldots,\, n_k. \tag{49}$$

Then for every $K \in \mathbb{N}$

$$(\mathcal{L})\int\int_{I_0} L_k d\mu \le \sum_{i=1}^{n_k} F(x_i^k,\, y_i^k)\mu(I_i^k) \le (\mathcal{L})\int\int_{I_0} U_k d\mu. \tag{50}$$

As $k \to \infty$,

$$\sum_{i=1}^{n_k} F(x_i^k,\, y_i^k)\mu(I_i^k) \to (\mathcal{R})\int\int_{I_0} F d\mu. \tag{51}$$

On the other hand, as $k \to \infty$

$$U_k \to F,\ L_k \to F \tag{52}$$

uniformly. This implies that

$$\lim_{k\to\infty}(\mathcal{L})\int\int_{I_0} L_k d\mu = \lim_{k\to\infty}(\mathcal{L})\int\int_{I_0} U_k d\mu = \int\int_{I_0} F d\mu. \tag{53}$$

(47) follows now easily from (50), (52), and (53).

2.4. Proposition. *Let $f \in \mathcal{H}$ and $\mu \in \mathcal{M}$. Then*

$$\int\int_{I_0} f(x,\, y)g_\mu(x,\, y)dxdy =$$

$$\tilde{f}(1,\, 1)g_\mu(1,\, 1) - \int_0^1 \tilde{f}(t,\, 1)dg_\mu(t,\, 1) - \int_0^1 \tilde{f}(1,\, t)dg_\mu(1,\, t) + \int\int_{I_0} \tilde{f} d\mu. \tag{54}$$

Proof. Since g_μ is of strongly bounded variation

$$\int\int_{I_0} f(x,\, y)g_\mu(x,\, y)dxdy \tag{55}$$

exists. Also, $\tilde{f} \in \mathcal{C}_0$ so that

$$\int\int_{I_0} \tilde{f} dg_\mu = (\mathcal{R})\int\int_{I_0} \tilde{f} d\mu \tag{56}$$

exists. The formula (54) follows now from the following integration by parts formula proved by Kurzweil in [4]:

$$\iint_{I_0} f(x, y)g_\mu(x, y)dxdy =$$

$$\iint_{I_0} \tilde{f}(x, y)dg_\mu(x, y) - \int_0^1 \tilde{f}(t, 1)dg_\mu(t, 1)$$

$$+\int_0^1 \tilde{f}(t, 0)dg_\mu(t, 0) - \int_0^1 \tilde{f}(1, t)dg_\mu(1, t) + \int_0^1 \tilde{f}(0, t)dg_\mu(0, t) \qquad (57)$$

$$+\tilde{f}(1, 1)g_\mu(1, 1) - \tilde{f}(1, 0)g_\mu(1, 0) - \tilde{f}(0, 1)g_\mu(0, 1) - \tilde{f}(0, 0)g_\mu(0, 0).$$

Obviously, (57) combined with 2.11 yields 2.14.

2.15. Remark. For $\mu \in \mathcal{M}$ the expression

$$\iint_{I_0} f(x, y)g_\mu(x, y)dxdy \qquad (58)$$

is a continuous linear functional on $\mathcal{H}$. We do not know, however, if (58) gives the general form of a continuous linear functional on $\mathcal{H}$. As we stated in 2.1, [8] shows that the general form of a continuous linear functional on $\mathcal{H}$ is

$$\iint_{I_0} \tilde{f} d\mu \qquad (59)$$

where $\mu \in \mathcal{M}$. Proposition 2.14 suggests the hypothesis that (58) is in fact another general form of a continuous linear functional on $\mathcal{H}$. We were not able to either prove or disprove it.

Also the following two problems are very natural.

2.16. Problem. Given a function $g : I_0 \to \mathbb{R}$ of strongly bounded variation, is there a $\mu \in \mathcal{M}$ such that g is equivalent to g_μ?

2.17. Problem. Given a multiplier g for Henstock integrable functions, is there a $\mu \in \mathcal{M}$ such that g is equivalent to g_μ?

3.1. The Henstock integral may be defined in an abstract setting, as presented in [3] and [7] (chapter 1). The problem of characterizing the multipliers for Henstock integrable functions remains unanswered then. However, [1] contains an interesting theorem on that subject. We will show that the theorem generalizes to spaces equipped with derivation bases, in which one can define the abstract Henstock integral.

3.2. Definition. Let X be a nonempty set, and Ψ be a nonvoid class of its subsets. A nonempty class Δ contained in the powerset of $X \times \Psi$ will be termed a ***derivation base*** on X. One can take X to be $\mathbb{R}$, $\mathbb{R}^2$, $\mathbb{R}^m$, or a locally compact Hausdorff space (these are the only settings considered until now). We will follow the notation in [7].

A ***partition*** π is a finite class consisting of elements of $X \times \Psi$ such that $D = \{I \in \Psi : (x, I) \in \pi\}$ has exactly as many elements as π and its elements are nonoverlapping (in the sense specified for X, the definition for $\mathbb{R}^m$ is the obvious one). If the union of all elements of D equals $I_0 \in \Psi$ then we say that π is a ***partition of*** I_0.

A base Δ has the ***partitioning property*** if for every $I \in \Psi$ and every $\alpha \in \Delta$ there exists a partition $\pi \subset \alpha$ of I.

If $I_0 \in \Psi$, and $F : X \times \Psi \to \mathbb{R}$ then the ***Henstock integral*** of F with respect to Δ over I_0 is a number $(\Delta)\int_{I_0} F$ such that for every ε there exists an $\alpha \in \Delta$ such that for every partition $\pi \subset \alpha$ of I_0

$$|\sum_{(x,I)\in\pi} F(x, I) - (\Delta)\int_{I_0} f| \leq \varepsilon. \tag{60}$$

Usually, we consider functions $F : X \times \Psi \to \mathbb{R}$ of the form $F(x, I) = f(x)\lambda(I)$ where f; $I_0 \to \mathbb{R}$ and $\lambda : \psi \to \mathbb{R}$ is additive. In this case we will write $(\Delta)\int_{I_0} f d\lambda$ for the Henstock integral of F.

Two functions $F_1 : X \times \Psi \to \mathbb{R}$, $F_2 : X \times \Psi \to \mathbb{R}$ are ***variationally equivalent*** on $I_0 \in \Psi$ if for every $\varepsilon > 0$ there exists an $\alpha \in \Delta$ and a superadditive nonnegative $\Omega : \Psi \to \mathbb{R}$ such that $\Omega(I_0) \leq \varepsilon$, and for every $(x, I) \in \alpha$ with $x \in I_0$

$$|F_1(x, I) - F_2(x, I)| \leq \Omega(I). \tag{61}$$

It is well known (see, for example, chapter 1 of [7]) that F is Henstock integrable if and only if there exists an additive $H : \Psi \to \mathbb{R}$ variationally equivalent to F. In fact, the Henstock integral

$$H(I) = (\Delta)\int_I F \tag{62}$$

is the additive function equivalent to the integrand.

For $\alpha \in \Delta$ and $E \subset X$ we define

$$\alpha[E] = \{(x, I) \in \alpha : x \in E\}, \alpha(E) = \{(x, I) \in \alpha : I \subset E\}. \tag{63}$$

Also

$$\Delta[E] = \{\alpha[E] : \alpha \in \Delta\}, \Delta(E) = \{\alpha(E) : \alpha \in \Delta\}. \tag{64}$$

These are called ***sections*** of the elements α of the base Δ, and of the base itself.

Δ has a *σ-local character* if for every sequence $\{X_n\}_{n\in\mathbb{N}}$ of disjoint subsets of X, and for every sequence $\{\beta_n\}_{n\in\mathbb{N}}$ such that for every $n \in \mathbb{N}$, $\beta_n \in \Delta[X_n]$, there exists an $\alpha \in \Delta$ such that for every $n \in \mathbb{N}, \alpha[X_n] \subset \beta_n$.

3.3. Theorem. *Let Δ be a derivation base with the partitioning property which is also of σ-local character. Let $F: X \times \Psi \to \mathbb{R}$, $F(x, I) = f(x)\lambda(I)$, where $f: X \to \mathbb{R}$, and $\lambda: \Psi \to \mathbb{R}$ is additive, be Henstock integrable with respect to Δ on $I_0 \in \Psi$. Then for a $G: X \times \Psi \to \mathbb{R}$ of the form $G(x, I) = f(x)g(x)\lambda(I)$, where $g: X \to \mathbb{R}$, G is Henstock integrable on I_0 if and only if $K(x, I) = g(x)H(I)$, where*

$$H(I) = (\Delta)\int_I f d\lambda, \tag{65}$$

is Henstock integrable with respect to Δ on I_0.

Proof. Let $\varepsilon > 0$ be arbitrary. Since F is Henstock integrable with H being its indefinite Henstock integral, F and H are variationally equivalent. For every $n \in \mathbb{N}$ there exists an α_n such that for every $\pi \subset \alpha_n$, a partition of I_0,

$$\sum_{(x,I)\in\pi} |f(x)\lambda(I) - \int_I f d\lambda| < \frac{\varepsilon}{n2^n}. \tag{66}$$

Let

$$E_n = \{x \in I_0 : (n-1) < |g(x)| \leq n\}, \tag{67}$$

for $n \in \mathbb{N}$. Then

$$I_0 = \bigcup_{n\in\mathbb{N}} E_n. \tag{68}$$

Since Δ has a σ-local character, there exists an $\alpha \in \Delta$ such that for every $n \in \mathbb{N}$

$$\alpha[E_n] \subset \alpha_n. \tag{69}$$

If $\pi \subset \alpha$ is a partition of I_0 then

$$|\sum_{(x,I)\in\pi} f(x)g(x)\lambda(I) - \sum_{(x,I)\in\pi} g(x)(\Delta)\int_I f d\lambda| \leq$$

$$|\sum_{(x,I)\in\pi} (f(x)g(x)\lambda(I) - g(x)(\Delta)\int_I f d\lambda)| \leq$$

$$\sum_{(x,I)\in\pi} |g(x)||f(x)\lambda(I) - \int_I f d\lambda| \leq \sum_{n=1}^{+\infty} \sum_{(x,I)\in\alpha_n\cap\pi} |g(x)||f(x)\lambda(I) - \int_I f d\lambda| \leq$$

$$\sum_{n=1}^{+\infty} n \sum_{(x,I)\in\alpha_n\cap\pi} |f(x)\lambda(I) - \int_I f d\lambda| \leq \sum_{n=1}^{+\infty} n\frac{\varepsilon}{n2^n} = \varepsilon. \tag{70}$$

This implies that G and K are variationally equivalent, and the theorem follows.

3.4. Corollary. *Let $f \in \mathcal{H}$ on I_0 and $g : I_0 \to \mathbb{R}$. Then $fg \in \mathcal{H}$ if and only if g is Henstock-integrable with respect to $\tilde{f}$ and*

$$\int\int_{I_0} f(x, y)g(x, y)dxdy = \int\int_{I_0} g(x, y)d\tilde{f}(x, y). \tag{71}$$

Proof. Notice that if

$$H(I) = \int\int_I f(x, y)dxdy \tag{72}$$

then for $I = [a, b] \times [c, d]$

$$H(I) = \tilde{f}(b, d) - \tilde{f}(a, c) - \tilde{f}(b, c) + \tilde{f}(a, c), \tag{73}$$

so that

$$\int\int_{I_0} gdH = \int\int_{I_0} g(x, y)d\tilde{f}(x, y). \tag{74}$$

The rest follows now from theorem **3.3.**

3.5. Corollary. *If $f \in \mathcal{H}$ and $\mu \in \mathcal{M}$ then g_μ is Henstock integrable with respect to $\tilde{f}$ and*

$$\int\int_{I_0} g_\mu d\tilde{f}(x, y) = \tag{75}$$

$$\int\int_{I_0} \tilde{f}d\mu - \int_0^1 \tilde{f}(t, 1)dg_\mu(t, 1) - \int_0^1 \tilde{f}(1, t)dg_\mu(1, t) + \tilde{f}(1, 1)g_\mu(1, 1).$$

Proof. This is another form of the integration by parts formula and it follows directly from corollary 3.4 and proposition 2.14.

References:

1. S.I. Ahmed, and W. F. Pfeffer, A Riemann integral in a locally compact Hausdorff space, *J. Austral. Math. Soc.*, **41A** (1986), 115-137.

2. R. Henstock, *Theory of integration*, Butterworths, 1963.

3. R. Henstock, Integration, variation, and differentiation in division spaces, *Proc. Royal Irish Acad.*, **78A** (1978), 69-85.

4. J. Kurzweil, On multiplication of Perron-integrable functions, *Czech. Math. J.*, **23**(98)(1973), 542-566.

5. J. Kurzweil, *Nichtabsolut konvergente Integrale,* Teubner Texte zür Mathematik, No. 26, Leipzig, 1980.

6. K. Ostaszewski, A topology for the spaces of Denjoy integrable functions, Proceedings of the Sixth Summer Real Analysis Symposium, *Real Analysis Exchange,* **9**(1) (1983-84), 79-85.

7. K. Ostaszewski, *Henstock Integration in the Plane,* Memoirs of the Amer. Math. Soc., (63)**353**, September 1986.

8. K. Ostaszewski, The space of Henstock integrable functions of two variables, *Internat. J. Math. and Math. Sci.,* (11)**1**(1988), 15-22.

9. B.S. Thomson, Spaces of conditionally integrable functions, *J. London Math. Soc.,* (2)**2**(1970), 358-360.

Piotr Mikusiński, Department of Mathematics, University of Central Florida, FL 32816

Krzysztof Ostaszewski[1], Department of Mathematics, University of Louisville Louisville, KY 40292

[1]**This author was partially supported by a University of Louisville research grant.**

Divergence theorem for vector fields with singularities

Washek F. Pfeffer

1. INTRODUCTION. We shall establish the divergence theorem in the m-dimensional Euclidean space for bounded vector fields which are continuous outside a set of $(m-1)$-dimensional Hausdorff measure zero, and differentiable (*not necessarily continuously*) outside a set of σ-finite $(m-1)$-dimensional Hausdorff measure. *No topological restrictions*, such as compactness, shall be imposed on the exceptional sets (cf. [P1] and [PY]). The examples of a smooth function with a single gap and the Cantor function (see [Ha, Section 19, Exercise (3), p.83]) show that, in terms of $(m-1)$-dimensional Hausdorff measure, the exceptional sets are as large as one can hope for.

As the divergence of a *noncontinuously* differentiable vector field need not be Lebesgue integrable, the full-fledged divergence theorem must be formulated by means of a more general integral than that of Lebesgue (cf. [Sh]). Thus we employ the variational approach of R. Henstock (see [He, Section 5, p. 54]) to define a coordinate free nonabsolutely convergent integral capable of integrating the divergence of a vector field with many singularities. As the integral is invariant with respect to smooth changes of coordinates, it is routine to lift it to a differentiable manifold and prove the Stokes theorem. The integral is also free of deficiencies displayed by the integrals defined in [JK1] and [JK2].

A brief outline (without proofs) of the results presented here can be found in [P2].

2. PRELIMINARIES. By $\mathbb{R}$ and $\mathbb{R}_+$ we denote the set of all real and all positive real numbers, respectively. All functions in this paper are real valued, and often the same letter is used to denote a function on a set A as well as its restriction to a set $B \subset A$. The algebraic operations, partial ordering, and convergence among functions on the same set are defined pointwise.

Throughout, $m \geq 1$ is a fixed integer, and $\mathbb{R}^m$ denotes the m-dimensional Euclidean space. For $x = (\xi_1, \ldots, \xi_m)$ and $y = (\eta_1, \ldots, \eta_m)$ in $\mathbb{R}^m$, we let $x \cdot y = \sum_{i=1}^m \xi_i \eta_i$, and set $||x|| = \sqrt{x \cdot x}$ and $|x| = \max(|\xi_1|, \ldots, |\xi_m|)$. Unless specified otherwise, in $\mathbb{R}^m$ we use exclusively the metric induced by the norm $|x|$. The distance between a point $x \in \mathbb{R}^m$ and a set $E \subset \mathbb{R}^m$ is denoted by $\operatorname{dist}(x, E)$. If $E \subset \mathbb{R}^m$, then E^-, E°, $E^\cdot$, and $d(E)$ denote, respectively, the closure, interior, boundary, and diameter of E; moreover, if $\delta > 0$, we set $U(E,\delta) = \{x \in \mathbb{R}^m : \operatorname{dist}(x, E) < \delta\}$. For each $x \in \mathbb{R}^m$, we write $U(x, \delta)$ instead of $U(\{x\}, \delta)$.

An interval $\Pi_{i=1}^m [k_i 2^{-n}, (k_i+1)2^{-n})$, where $n \geq 0$ and $k_1, \ldots, k_m$ are integers, is called a ***dyadic cube***. Often we shall use the simple observation that any family of dyadic cubes contains a disjoint subfamily which has the same union as the original family.

By $\mathcal{H}$ we denote the $(m-1)$-dimensional outer Hausdorff measure in $\mathbb{R}^m$ as defined in [Fe, Section 2.10.2, p.171]. If $k \geq 1$ is an integer, then λ_k denotes the k-dimensional outer Lebesgue measure in $\mathbb{R}^k$. We write λ instead of λ_1, and $|E|$ instead of $\lambda_m(E)$ for each $E \subset \mathbb{R}^m$. The words "outer measure", "measure", and "measurable", as well as the expression "almost all", always refer to λ_m.

Note that the measure $\mathcal{H}$ is defined so that $\mathcal{H}(E) = \lambda_{m-1}(E)$ for each set $E \subset \mathbb{R}^{m-1}$. In particular, $\mathcal{H}$ is a constant multiple (by a constant different from one) of the measure $\mathcal{H}^{m-1}$ defined in [Fa, Section 1.2, p.7] — cf. [Fa, Theorem 1.12, p.13].

A subset of $\mathbb{R}^m$ is called ***slight*** or ***thin***, respectively, according to whether its outer measure $\mathcal{H}$ is zero or σ-finite. We note that the slight and thin sets defined here are appreciably ***larger*** than those defined in [P2] and [PY]; in particular, they need not be compact. Clearly, $|E| = 0$ for each thin set $E \subset \mathbb{R}^m$.

A bounded set $A \subset \mathbb{R}^m$ is called ***admissible*** if $A^{\cdot}$ is thin, and the distributional gradient of the characteristic function of A is a vector measure on Borel subsets of $\mathbb{R}^m$ whose variation is finite (see[M, Definition 3]). In particular, admissible sets are ***Caccioppoli sets*** defined in [G, Definition 1.7, p.6]. According to [Fe, Chapter 4], to each admissible set A we can associate a vector field n_A (usually referred to as the ***Federer exterior normal***) such that

$$\int_A \operatorname{div} v \, d\lambda = \int_{A^{\cdot}} v \cdot n_A \, d\mathcal{H}$$

for each vector field v continuously differentiable in a neighborhood of A^-.

Let $\mathcal{A}$ be the family of all admissible sets. For $A \in \mathcal{A}$, we set $\|A\| = \int_{A^{\cdot}} \|n_A\| d\mathcal{H}$, and let

$$r(A) = \begin{cases} \frac{|A|}{d(A)\|A\|} & \text{if } d(A)\|A\| > 0 \\ 0 & \text{otherwise.} \end{cases}$$

The numbers $\|A\|$ and $r(A)$ are called the ***perimeter*** and ***regularity*** of A, respectively. If $E \subset \mathbb{R}^m$, we let

$$\mathcal{A}(E) = \{B \in \mathcal{A} : B \subset E\} \text{ and } \mathcal{A}_\circ(E) = \{B \in \mathcal{A} : B^- \subset E\}.$$

2.1. Lemma. There is a positive constant α, depending only on m, with the following property: for every set $E \subset \mathbb{R}^m$ with $\mathcal{H}(E) < +\infty$, and for each $\eta > 0$, there is a countable disjoint family $\mathcal{C}$ of dyadic cubes such that $E \subset (\bigcup \mathcal{C})^\circ, |\bigcup \mathcal{C}| < \eta$, and

$$\sum_{C \in \mathcal{C}} ||C|| < \alpha \mathcal{H}(E) + \eta.$$

Proof. It follows from [Fa, Theorem 5.1, p.65] that there is a constant $a > 0$, depending only on m, such that for each $\delta > 0$, we can find a countable family $\mathcal{K}$ of dyadic cubes such that $E \subset (\bigcup \mathcal{K})^\circ$,

$$\sum_{k \in \mathcal{K}} [d(K)]^{m-1} < a\mathcal{H}(E) + \frac{\eta}{2m},$$

and $d(K) < \delta$ for each $K \in \mathcal{K}$. Let $\alpha = 2ma$, and choose $\delta \in (0,1)$ so that $a\mathcal{H}(E)\delta < \eta/2$. Then

$$\sum_{k \in \mathcal{K}} ||K|| = 2m \sum_{K \in \mathcal{K}} [d(K)]^{m-1} < \alpha \mathcal{H}(E) + \eta$$

and

$$|\bigcup \mathcal{K}| \le \sum_{K \in \mathcal{K}} |K| \le \delta \sum_{K \in \mathcal{K}} [d(K)]^{m-1} < \delta[a\mathcal{H}(E) + \frac{\eta}{2m}] < \eta.$$

Now it suffices to select a disjoint family $\mathcal{C} \subset \mathcal{K}$ with $\bigcup \mathcal{C} = \bigcup \mathcal{K}$.

3. ADDITIVE FUNCTIONS. A ***division*** of an admissible set A is a finite disjoint family $\mathcal{D} \subset \mathcal{A}$ with $\bigcup \mathcal{D} = A$. An ***additive*** function in an admissible set A is a function F defined on $\mathcal{A}(A)$ and such that $F(A) = \sum_{D \in \mathcal{D}} F(D)$ for each division $\mathcal{D}$ of A. It is easy to show that if F is an additive function in an admissible set A and $B \in \mathcal{A}(A)$, then $F \restriction \mathcal{A}(B)$ is an additive function in B.

3.1 Definition. Let F be an additive function in an admissible set A. We say that F is:

1. ***lower bounded*** if given $\varepsilon > 0$, there is a $\delta > 0$ such that $F(B) > -\varepsilon$ for each admissible set $B \subset A$ with $||B|| < \delta$;

2. ***lower continuous*** in a set $E \subset A^-$ if given $\varepsilon > 0$ there is a $\delta > 0$ such that $F(B) > -\varepsilon$ for each admissible set $B \subset A \cap U(E, \delta)$ with $|B| < \delta$ and $||B|| < 1/\varepsilon$.

If both F and $-F$ are lower bounded or lower continuous, we say that F is ***bounded*** or ***continuous***, respectively.

3.2. Remark. Let F be an additive function in an admissible set A. Clearly, for each $E \subset A^-$ the lower continuity of F in E is equivalent to that in E^-. Moreover, it follows from the isoperimetric inequality (1.19) in [G, Corollary 1.29, p. 25] that if F is lower continuous in A, then it is lower bounded.

3.3. Example. Let $A \in \mathcal{A}$ and let f be a measurable function on A with $\int_A |f| d\lambda_m < +\infty$. If $F(B) = \int_B f d\lambda_m$ for each admissible set $B \subset A$, then F is an additive function in A, which is continuous in A by the absolute continuity of the indefinite Lebesgue integral.

3.4. Example. Let $A \in \mathcal{A}$ and let v be a bounded $\mathcal{H}$-measurable vector field on A^-, which is continuous in a ***compact*** set $E \subset A^-$. If $F(B) = \int_{B^\cdot} v \cdot n_B d\mathcal{H}$ for each admissible set $B \subset A$, then F is a bounded additive function in A, which is continuous in E. The additivity of F is obvious, and since

$$|F(B)| \leq \int_{B^\cdot} |v \cdot n_B| d\mathcal{H} \leq ||B|| \cdot \sup\{v(x) : x \in A^-\},$$

we see that F is bounded. To show that F is continuous in E requires a proof.

Choose an $\varepsilon > 0$, and using the Stone-Weierstrass theorem, find a vector field w with polynomial coordinates and such that $||v(x) - w(x)|| < \varepsilon^2/6$ for each $x \in E$. For every $x \in E$, find a $\delta_x > 0$ so that

$$||v(y) - v(x)|| < \varepsilon^2/6 \text{ and } ||w(y) - w(x)|| < \varepsilon^2/6$$

whenever $y \in A^- \cap U(x, \delta_x)$. By the compactness of E, there is a $\delta > 0$ such that $U(E, \delta) \subset \bigcup_{x \in E} U(x, \delta_x)$. If $y \in U(E, \delta)$, then $y \in U(x, \delta_x)$ for some $x \in E$, and we have

$$||v(y) - w(y)|| \leq ||v(y) - v(x)|| + ||v(x) - w(x)|| + ||w(x) - w(y)|| < \varepsilon^2/2.$$

By making δ smaller, we may assume that $\gamma\delta < \varepsilon/2$ where

$$\gamma = \sup\{|\text{div } w(x)| : x \in A^-\}.$$

Now if $B \subset A \cap U(E, \delta)$ is an admissible set with $|B| < \delta$ and $||B|| < 1/\varepsilon$, then

$$|F(B)| \leq |\int_{B^\cdot} (v - w) \cdot n_B d\mathcal{H}| + |\int_{B^\cdot} w \cdot n_B d\mathcal{H}|$$

$$\leq \frac{\varepsilon^2}{2} \int_{B^\cdot} ||n_B|| d\mathcal{H} + \int_B |\text{div } w| d\lambda_m \leq \frac{\varepsilon^2 ||B||}{2} + \gamma|B| < \varepsilon,$$

and the continuity of F in E is established.

If A and B are subsets of $\mathbb{R}^m$, then the number

$$A[B] = \inf\{\delta > 0 : B \subset U(A, \delta)\}$$

measures the extent to which B is ***not contained*** in A. We shall use this concept to define a convergence in the family of admissible sets which is related to the continuity of additive functions. Let $E \subset \mathbb{R}^m$, $B \in \mathcal{A}$, and let $\{B_n\}$ be a sequence in $\mathcal{A}$. We say that

1. $\{B_n\}$ ***converges*** to B, and write $\{B_n\} \to B$, if $B_n \subset B$, $n = 1, 2, \ldots$, and $\lim ||B - B_n|| = 0$;

2. $\{B_n\}$ *E-converges* to B, and write $\{B_n\} \xrightarrow{E} B$, if $B_n \subset B$, $n = 1, 2, \ldots$, $\sup ||B_n|| < +\infty$, $\lim |B - B_n| = 0$, and $\lim E[B - B_n] = 0$.

If $\{B_n\} \to B$, then it follows from the isoperimetric inequality (1.19) in [G, Corollary 1.29, p.25] that $\{B_n\} \xrightarrow{E} B$ whenever $B \subset E^-$.

The next lemma characterizes the lower boundedness and lower continuity of additive functions in terms of convergence.

3.5 Lemma. Let $A \in \mathcal{A}$, $E \subset A^-$, and let F be an additive function in A.

1. F is lower bounded if and only if $F(B) \geq \limsup F(B_n)$ for each sequence $\{B_n\}$ in $\mathcal{A}(A)$ which converges to $B \in \mathcal{A}(A)$.

2. F is lower continuous in E if and only if $F(B) \geq \limsup F(B_n)$ for each sequence $\{B_n\}$ in $\mathcal{A}(A)$ which E-converges to $B \in \mathcal{A}(A)$

Proof. We only prove 2; the proof of 1 is analogous and simpler. Let F be lower continuous in E, $B \in \mathcal{A}(A)$, and let $\{B_n\}$ be a sequence in $\mathcal{A}(A)$ which E-converges to B. Choose an $\varepsilon > 0$ with $1/\varepsilon > ||B|| + \sup ||B_n||$, and find a $\delta > 0$ so the $F(C) > -\varepsilon$ for each admissible set $C \subset A \cap U(E, \delta)$ with $|C| < \delta$ and $||C|| < 1/\varepsilon$. There is an integer $N \geq 1$ such that $|B - B_n| < \delta$ and $E[B - B_n] < \delta$ for each $n \geq N$. Thus $B - B_n \subset U(E, \delta)$ for $n = N, N+1, \ldots$, and by [M, Theorem 35],

$$||B - B_n|| \leq ||B|| + ||B_n|| < 1/\varepsilon$$

for $n = 1, 2, \ldots$. As F is additive, we have

$$F(B_n) = F(B) - F(B - B_n) < F(B) + \varepsilon$$

for each $n \geq N$. Consequently $\limsup F(B_n) \leq F(B) + \varepsilon$, and from the arbitrariness of ε we conclude that $\limsup F(B_n) \leq F(B)$.

Conversely, if F is not lower continuous in E, then there is an $\varepsilon > 0$ and a sequence $\{B_n\}$ of admissible subsets of $A \cap U(E, 1/n)$ such that $|B_n| < 1/n$, $||B_n|| < 1/\varepsilon$, and

$$-\varepsilon \geq F(B_n) = F(A) - F(A - B_n)$$

for $n = 1, 2, \ldots$. It follows that $\limsup F(A - B_n) > F(A)$, and yet, using [M, Theorem 35], it is easy to see that $\{A - B_n\} \xrightarrow{E} A$.

3.6 Definition. Let F be an additive function in an admissible set A. We say that F is *lower amiable* if it is lower bounded, and if there is a slight set $S \subset A^-$ such that F is lower continuous in each compact set $E \subset A^- - S$. If both F and $-F$ are lower amiable, we say that F is *amiable*.

3.7 Lemma. Let F be a lower amiable function in an admissible set A, and let T be a compact thin subset of A^-. Then for each $\varepsilon > 0$ there is a set K which is a finite union of dyadic cubes, and such that $T \subset K^\circ$ and $F(A \cap K) > -\varepsilon$.

Proof. If L is a dyadic cube, we denote by $\mathcal{L}^*$ the collection of all dyadic cubes of diameter $d(L)$ which are adjacent to L, and set $L^* = \bigcup \mathcal{L}^*$. Choose an $\varepsilon > 0$, and fix an integer $k \geq 1$. Let $S \subset A^-$ be the slight set associated to F by Definition 3.6. Using Lemma 2.1, find a sequence $\{S_n\}$ of dyadic cubes so that S is contained in $G = (\bigcup_n S_n)^\circ$ and $\sum_n ||S_n|| < 1/k$. As F is lower continuous in $A^- - G$, for

$$\varepsilon_i = \min[2^{-i-1}\varepsilon,\ (1 + ||A||)^{-1}],$$

$i = 1, 2, \ldots$, there is a $\delta_i > 0$ such that $F(B) > -\varepsilon_i$ whenever B is an admissible subset of $U(A^- - G,\ \delta_i)$ with $|B| < \delta_i$ and $||B|| < 1/\varepsilon_i$.

Let α be a positive constant from Lemma 2.1. We may assume that $T - G = \bigcup_{i=1}^{\infty} T_i$ where $\mathcal{H}(T_i) < 3^{-m}/\alpha$ for $i = 1, 2, \ldots$. By Lemma 2.1, there are sequences $\{T_{i,n}\}_n$ of dyadic cubes such that $d(T_{i,n}) < \delta_i$ for $n = 1, 2, \ldots$, $T_i \subset \bigcup_n T_{i,n}$, and $\sum_n ||T_{i,n}|| < 3^{-m}$. The family $\{(T_{i,n}^*)^\circ : i, n = 1, 2, \ldots\}$ is an open cover of the compact set $T - G$, and so

$$T - G \subset [\bigcup_{i=1}^{r} (\bigcup_{n=1}^{p_i} T_{i,n}^*)]^\circ$$

for some positive integers r and p_i, $i = 1, \ldots, r$. If $\mathcal{C}$ is the collection of those cubes from $\bigcup_{i=1}^{r} (\bigcup_{n=1}^{p_i} T_{i,n}^*)$ whose closures meet $T - G$, then $T - G$ is contained in $H = (\bigcup \mathcal{C})^\circ$. By eliminating the superfluous cubes, we may assume that $\mathcal{C}$ is a disjoint family. For $i = 1, \ldots, r$ let

$$\mathcal{C}_i = \mathcal{C} \cap (\bigcup_{n=1}^{p_i} T_{i,n}^*) \text{ and } C_i = \bigcup (\mathcal{C} - \bigcup_{j=1}^{i-1} \mathcal{C}_j).$$

Then each C_i is a finite union of dyadic subcubes of $U(T - G, \delta_i)$, and $C_1, \ldots, C_r$ are disjoint sets whose union is equal to the union of $\mathcal{C}$. We have

$$|C_i| \leq 3^m \sum_{n=1}^{p_i} |T_{i,n}| \leq 3^m \delta_i \sum_{n=1}^{p_i} ||T_{i,n}|| < \delta_i,$$

and by [M, Theorem 35],

$$||C_i|| \leq 3^m \sum_{n=1}^{p_i} ||T_{i,n}|| < 1.$$

As T is covered by H and the interiors of the sets S_n^*, $n = 1, 2, \ldots$, there is a positive integer p such that

$$T \subset H \cup (\bigcup_{n=1}^{p} (S_n^*)^\circ) \subset [(\bigcup_{i=1}^{r} C_i) \cup (\bigcup_{n=1}^{p} S_n^*)]^\circ.$$

If D is the union of all cubes from $\bigcup_{n=1}^{p} \mathcal{S}_n^*$ which do not meet H, then $T \subset [(\bigcup_{i=1}^{r} C_i) \cup D]^\circ$, and by [M, Theorem 35],

$$||D|| \leq 3^m \sum_{n=1}^{p} ||S_n|| < 3^m/k.$$

Since $A \cap C_i$ is contained in $U(A^{\cdot} - G, \delta_i)$, $|A \cap C_i| < \delta_i$, and by [M, Theorem 35],

$$||A \cap C_i|| \le ||A|| + ||C_i|| < ||A|| + 1 \le 1/\varepsilon_i,$$

we have

$$\sum_{i=1}^{r} F(A \cap D_i) > -\sum_{i=1}^{r} \varepsilon_i > -\varepsilon/2.$$

As $||D|| > 3^m/k$, it follows from [MM, Section 13] that by taking k sufficiently large, we can make $||A \cap D||$ so small that $F(A \cap D) > -\varepsilon/2$; for F is lower bounded. Thus

$$\sum_{i=1}^{r} F(A \cap C_i) + F(A \cap D) > -\varepsilon,$$

and it suffices to let $K = (\bigcup_{i=1}^{r} C_i) \cup D$.

The next lemma is proved by a technique of A.S.Besicovitch introduced in [B].

3.8 Lemma. Let F be an amiable function in a dyadic cube L, and let T be a thin subset of L^-. If $F(L) < 0$, then there is an $x \in L^- - T$ and a strictly decreasing sequence $\{L_n\}$ of dyadic subcubes of L such that $\bigcap_{n=1}^{\infty} L_n = \{x\}$ and $F(L_n) < 0$ for $n = 1, 2, \ldots$.

Proof. Throughout this proof only, for a set $E \subset L^-$, we define E° as the interior of E *relative* to L^-. If K is a dyadic subcube of L, we denote by $\mathcal{K}^*$ the collection of all dyadic cubes $C \subset L$ of diameter $d(K)$ which are adjacent to K, and set $K^* = \bigcup \mathcal{K}^*$.

Let $S \subset L^-$ be the slight set associated to F by Definition 3.6. Since F is lower bounded, we can find a $\delta > 0$ so that $F(B) > F(L)/2$ for each $B \in \mathcal{A}(L)$ with $||B|| < 3^m\delta$. In view of Lemma 2.1, there is a sequence $\{S_n\}$ of dyadic subcubes of L such that $S \subset \bigcup_n S_n^-$ and $\sum_n ||S_n|| < \delta$. The set S is contained in $G = \bigcup_n (S_n^{*-})^\circ$, and if $\mathcal{S} = \bigcup_n \mathcal{S}_n^*$, then

$$\sum_{K \in \mathcal{S}} ||K|| \le 3^m \sum_n ||S_n|| < 3^m \delta.$$

Thus if $\{K_1, \ldots, K_p\}$ is a disjoint subfamily of $\mathcal{S}$, then

$$\sum_{n=1}^{p} F(K_n) = F(\bigcup_{n=1}^{p} K_n) > F(L)/2;$$

for by [M, Theorem 35],

$$||\bigcup_{n=1}^{p} K_n|| \le \sum_{n=1}^{p} ||K_n|| < 3^m \delta.$$

In particular, L does not belong to $\mathcal{S}$.

As F is lower continuous in $L^- - G$, for $\varepsilon_i = \min[-2^{-i-1}F(L), 1]$, $i = 1, 2, \ldots$, there is a $\delta_i > 0$ such that $F(B) > -\varepsilon_i$ whenever B is an admissible subset of $U(L^- - G, 2\delta_i)$ with $|B| < \delta_i$ and $||B|| < 1/\varepsilon_i$. Let α be the positive constant from Lemma 2.1. We may assume that $T - G = \bigcup_{i=1}^{\infty} T_i$ where $\mathcal{H}(T_i) < 3^m/\alpha$ for $i = 1, 2, \ldots$.

Fix an integer $i \geq 1$. By Lemma 2.1, there is a sequence $\{T_{i,n}\}_n$ of dyadic subcubes of L such that $d(T_{i,n}) < \delta_i$ and $T_i \cap T_{i,n}^- \neq \emptyset$ for $n = 1, 2, \ldots$, $T_i \subset \bigcup_n T_{i,n}^-$, and $\sum_n ||T_{i,n}|| < 3^{-m}$. It is easy to see that $T_i \subset \bigcup_n (T_{i,n}^{*-})^\circ$ and

$$T_{i,n}^* \subset U(T_i, 2\delta_i) \subset U(L^- - G, 2\delta_i).$$

If $\mathcal{T}_i = \bigcup_n \mathcal{T}_{i,n}^*$, then

$$\sum_{K \in \mathcal{T}_i} ||K|| \leq 3^m \sum_n ||T_{i,n}|| < 1$$

and

$$\sum_{K \in \mathcal{T}_i} |K| \leq 3^m \sum_n |T_{i,n}| \leq 3^m \delta_i \sum_n ||T_{i,n}|| < \delta_i.$$

Thus if $\{K_1, \ldots, K_p\}$ is a disjoint subfamily of $\mathcal{T}_i$, then

$$\sum_{n=1}^{p} F(K_n) = F(\bigcup_{n=1}^{p} K_n) > -\varepsilon_i;$$

for $\bigcup_{n=1}^{p} K_n \subset U(L^- - G, 2\delta_i)$,

$$|\bigcup_{n=1}^{p} K_n| \leq \sum_{n=1}^{p} |K_n| < \delta_i,$$

and by [M,Theorem 35],

$$||\bigcup_{n=1}^{p} K_n|| \leq \sum_{n=1}^{p} ||K_n|| < 1 \leq 1/\varepsilon_i.$$

In particular, L does not belong to $\mathcal{T}_i$.

We denote $\mathcal{S} \cup (\bigcup_{i=1}^{\infty} \mathcal{T}_i)$ by $\mathcal{K}$. For $k = 1, 2, \ldots$, let $\mathcal{C}_k$ be the collection of all the S_n's and $T_{i,n}$'s of diameter $d(L)/2^k$, and let $C_k = \bigcup\{C^* : C \in \mathcal{C}_k\}$. We have

$$T \subset G \cup (\bigcup_{i=1}^{\infty} T_i) \subset (\bigcup_n (S_n^{*-})^\circ) \cup (\bigcup_{i=1}^{\infty} \bigcup_n (T_{i,n}^{*-})^\circ) =$$

$$\bigcup_{k=1}^{\infty} \bigcup\{(C^{*-})^\circ : C \in \mathcal{C}_k\} \subset \bigcup_{k=1}^{\infty} [\bigcup\{C^{*-} : C \in \mathcal{C}_k\}]^\circ \subset \bigcup_{k=1}^{\infty} (C_k^-)^\circ.$$

Let $\mathcal{K}_k$ be the family of all dyadic cubes $K \subset C_k$ of diameter $d(L)/2^k$. Clearly $\mathcal{K}_k$ is a division of C_k, and $\bigcup_{k=1}^{\infty} \mathcal{K}_k = \mathcal{K}$. Combining the results of the previous paragraphs, we obtain that

$$\sum_{E \in \mathcal{K}'} F(E) = \sum\{F(E) : E \in \mathcal{K}' \cap \mathcal{S}\} +$$

$$\sum_{i=1}^{\infty}\left(\sum\{F(E):\ E\in(\mathcal{K}'-\mathcal{S})\cap(\mathcal{T}_i-\cup_{j=1}^{i-1}\mathcal{T}_j\}\right)$$

$$>\frac{F(L)}{2}-\sum_{i=1}^{\infty}\varepsilon_i\geq F(L)$$

for each finite disjoint family $\mathcal{K}'\subset\mathcal{K}$.

We shall construct inductively a decreasing sequence $\{L_k\}$ of dyadic subcubes of L so that $C_k\cap L_k=\emptyset$, $d(L_k)=d(L)/2^k$, and

$$F(L_k)<\min(0,\sum_{K\in\mathcal{K}'}F(K))$$

for each finite disjoint family $\mathcal{K}'\subset\mathcal{K}$ with $\cup\mathcal{K}'\subset L_k$. Let $\mathcal{D}_1$ be the family of all dyadic cubes $D\subset L-C_1$ of diameter $d(L)/2$, and suppose that for each $D\in\mathcal{D}_1$, either $F(D)\geq 0$, or $F(D)\geq\sum\{F(K):\ K\in\mathcal{K}_D\}$ for some finite disjoint family $\mathcal{K}_D\subset\mathcal{K}$ with $\cup\mathcal{K}_D\subset D$. Since $\mathcal{K}_1\cup\mathcal{D}_1$ is a division of L, and $\mathcal{K}_1\cup(\cup\{\mathcal{K}_D:\ D\in\mathcal{D}_1,\ F(D)<0\})$ is a finite disjoint subfamily of $\mathcal{K}$, we obtain

$$F(L)=\sum_{K\in\mathcal{K}_1}F(K)+\sum_{D\in\mathcal{D}_1}F(D)\geq\sum_{K\in\mathcal{K}_1}F(K)+$$

$$\sum\{F(D):\ D\in\mathcal{D}_1,\ F(D)<0\}\geq\sum_{K\in\mathcal{K}_1}F(K)+$$

$$\sum\{\sum_{K\in\mathcal{K}_D}F(K):\ D\in\mathcal{D}_1,\ F(D)<0\}>F(L),$$

which is a contradiction. Consequently, there is an $L_1\in\mathcal{D}_1$ having the desired properties. Suppose that for an integer $k\geq 2$, we have constructed a dyadic cube L_{k-1} possessing the desired properties. Let $\mathcal{D}_k$ be the family of all dyadic cubes $D\subset L_{k-1}-C_k$ of diameter $d(L)/2^k$, and suppose that for each $D\in\mathcal{D}_k$, either $F(D)\geq 0$, or $F(D)\geq\sum\{F(K):\ F\in\mathcal{K}_D\}$ for some finite disjoint family $\mathcal{K}_D\subset\mathcal{K}$ with $\bigcup\mathcal{K}_D\subset D$. If $\mathcal{K}'_k=\{K\in\mathcal{K}_k:\ K\subset L_{k-1}\}$, then $\mathcal{K}'_k\cup\mathcal{D}_k$ is a division of L_{k-1}, and $\mathcal{K}'_k\cup(\cup\{\mathcal{K}_D:\ D\in\mathcal{D}_k,\ F(D)<0\})$ is a finite disjoint subfamily of $\mathcal{K}$ whose union is contained in L_{k-1}. Thus by the induction hypothesis,

$$F(L_{k-1})=\sum_{K\in\mathcal{K}'_k}F(K)+\sum_{D\in\mathcal{D}_k}F(D)\geq\sum_{K\in\mathcal{K}'_k}F(K)+$$

$$\sum\{F(D):\ D\in\mathcal{D}_k,\ F(D)<0\}\geq\sum_{K\in\mathcal{K}'_k}F(K)+$$

$$\sum\{\sum_{K\in\mathcal{K}_D}F(K):\ D\in\mathcal{D}_k,\ F(D)<0\}>F(L_{k-1}),$$

which is a contradiction. Consequently, there is an $L_k\in\mathcal{D}_k$ having the desired properties.

Since C_k is a finite union of dyadic subcubes of L, and $C_k \cap L_k = \emptyset$, it is easy to verify the $[(C_k^-)^\circ] \cap (L_k^-) = \emptyset$. Thus if $\cap_{k=1}^\infty L_k^- = \{x\}$, then $x \notin T$, and the lemma is proved.

3.9 Proposition. Let F be a lower amiable function in an admissible set A, and let T be a thin subset of A^-. If $F(A) < 0$, then there is an $x \in A^\circ - T$, and a sequence $\{C_n\}$ of cubes in A with $\lim d(C_n) = 0$, and such that $x \in C_n^-$ and $F(C_n) < 0$ for $n = 1, 2, \ldots$.

Proof. By Lemma 3.7, we can find a set K which is a finite union of dyadic cubes, and such that $A^\cdot \subset K^\circ$ and $F(A \cap K) > F(A)$. Now there are disjoint dyadic cubes $L_1, \ldots L_n$ with $\cup_{i=1}^n L_i = A - K$, and by the additivity of F, we have

$$\sum_{i=1}^{n} F(L_i) = F(A) - F(A \cap K) < 0.$$

In view of Lemma 3.8, the proposition follows.

4. THE INTEGRAL. Let $A \in \mathcal{A}$, and let f and F be functions defined on A and $\mathcal{A}(A)$, respectively. Given an $\varepsilon > 0$ and a set $E \subset A^-$, an ***ε-majorant*** of the pair (f, F) in E is a nonnegative additive function M in A which satisfies the following conditions: $M(A) < \varepsilon$, and for every $x \in E$ there is a $\delta > 0$ such that

$$|f(x)|B| - F(B)| \leq M(B)$$

for each $B \in \mathcal{A}(A)$ with $x \in B^-$, $d(B) < \delta$, and $r(B) > \varepsilon$.

4.1 Definition. Let f be a function defined on an admissible set A. We say that f is ***integrable*** in A if we can find an amiable function F in A which satisfies the following condition: for each $\varepsilon > 0$ there is a thin set $T \subset A^-$ such that the pair (f, F) has an ε-majorant in $A^- - T$.

4.2 Remark. (i) As the boundaries of admissible sets are thin, in the previous difinition, we can always assume that the thin set T contains $A^\cdot$. Thus it suffices to consider ε-majorants on subsets of the interiors of admissible sets.

(ii) Since a countable union of thin sets is again thin, it is easy to see that the thin set T in the above definition can be selected independently of ε.

The family of all integrable functions in an admissible set A is denoted by $\mathcal{I}(A)$. If $f \in \mathcal{I}(A)$ for some admissible set A, then each amiable function F in A which satisfies the condition of Definition 4.1 is called ***an indefinite integral*** of f in A. Using the results of Section 3, we show that each integrable function has ***precisely one*** indefinite integral.

4.3 Lemma. Let $A \in \mathcal{A}$, and F_i be an indefinite integral of $f_i \in \mathcal{I}(A)$, $i = 1, 2$. If $f_1 \leq f_2$, then $F_1 \leq F_2$.

Proof. If $B \in \mathcal{A}(A)$, then clearly $F_i \lceil \mathcal{A}(B)$ is an indefinite integral of $f_i \lceil B$, $i = 1, 2$. Thus it suffices to show that the assumption $F_2(A) < F_1(A)$ leads to a contradiction. Choose an $\varepsilon > 0$ so that $\varepsilon < 1/2m$ and $F_2(A) + 2\varepsilon < F_1(A)$. For i = 1, 2, we can find thin sets $T_i \subset A^-$ and ε-majorants M_i of the pairs (f_i, F_i) in $A^- - T_i$. We see that $T = T_1 \cup T_2$ is a thin subset of A^-, and that

$$F = F_2 - F_1 + M_1 + M_2$$

is a lower amiable function in A with $F(A) < 0$. By Corollary 3.9, there is an $x \in A^\circ - T$, and a sequence $\{C_n\}$ of cubes in A with $\lim d(C_n) = 0$, and such that $x \in C_n^-$ and $F(C_n) < 0$ for $n = 1, 2, \ldots$. As $r(C_n) > \varepsilon$, it follows from Definition 4.1 that

$$|f_i(x)|C_N| - F_i(C_N)| \leq M_i(C_N)$$

for $i = 1, 2$ and some sufficiently large integer $N \geq 1$. This implies that

$$F_1(C_N) - M_1(C_N) \leq f_1(x)|C_N| \leq f_2(x)|C_N| \leq F_2(C_N) + M_2(C_N).$$

and hence $F(C_N) \geq 0$; a contradiction.

4.4 Corollary. If $A \in \mathcal{A}$ and $f \in \mathcal{I}(A)$, then all indefintie integrals of f in A are equal.

Let $A \in \mathcal{A}$ and $f \in \mathcal{I}(A)$. In view of Corollary 4.4, we can talk about ***the indefinite integral*** of f in A, denoted by $I(f, \cdot)$. The number $I(f, A)$ is called the ***integral*** of f over A.

4.5 Proposition. If $A \in \mathcal{A}$, then $\mathcal{I}(A)$ is a linear space, and the map $f \mapsto I(f, A)$ is a nonnegative linear functional on $\mathcal{I}(A)$.

Proof. If $f \in \mathcal{I}(A)$ and $f \geq 0$, then it follows from Lemma 4.3 that $I(f, A) \geq 0$. The rest of the proposition follows directly from Definition 4.1.

4.6 Proposition. Let $A \in \mathcal{A}$, $f : A \to \mathbb{R}$, and let $\mathcal{D}$ be a division of A. Then $f \in \mathcal{I}(A)$ if and only if $F\lceil D \in \mathcal{I}(D)$ for each $D \in \mathcal{D}$.

Proof. If $f \in \mathcal{I}(A)$, $F = I(f, \cdot)$, and $D \in \mathcal{D}$, then it is clear that $F\lceil\mathcal{A}(D)$ is the indefinite integral of $f\lceil D$ in D, and so $f\lceil D \in \mathcal{I}(D)$.
Conversely, suppose that $f\lceil D \in \mathcal{I}(D)$ for each $D \in \mathcal{D}$, and let

$$F_D = I(f\lceil D, \cdot).$$

For every $B \in \mathcal{A}(A)$, we set

$$F(B) = \sum_{D \in \mathcal{D}} F_D(B \cap D),$$

and we show that F is the indefinite integral of f in A. Using [M, Theorem 35] and [MM, Section 13], it is easy to verify that F is an amiable function in A. Let n be the number of elements of $\mathcal{D}$, and let $\varepsilon > 0$. For each $D \in \mathcal{D}$ find a thin set $T_D \subset D^-$ so that the pair $(f\lceil D, F_D)$ has an (ε/n)-majorant M_D in $D - T_D$. Let $T = \cup_{D \in \mathcal{D}}(D^\cdot \cup T_D)$, and set

$$M(B) = \sum_{D \in \mathcal{D}} M_D(B \cap D)$$

for every $B \in \mathcal{A}(A)$. Fix an $x \in A - T$. Then $x \in D^\circ$ for some $D \in \mathcal{D}$, and we can find $\eta > 0$ so that $E^- \subset D$ whenever $x \in E^-$ and $d(E) < \eta$. Moreover, there is a $\delta > 0$ such that

$$|f(x)|B| - F_D(B)| \leq M_D(B)$$

for each $B \in \mathcal{A}(D)$ with $x \in B^-$, $d(B) < \delta$, and $r(B) > \varepsilon/m$. Thus if $B \in \mathcal{A}(A)$, $x \in B^-$, $r(B) > \varepsilon \geq \varepsilon/m$, and $d(B) < \min(\eta, \delta)$, then $B \in \mathcal{A}(D)$ and we have

$$|f(x)|B| - F(B)| = |f(x)|B| - F_D(B)| \leq M_D(B) = M(B).$$

Since T is a thin set, and

$$M(A) = \sum_{D \in \mathcal{D}} M_D(D) < n \cdot (\varepsilon/n) = \varepsilon,$$

the proposition is proved.

If $E \subset \mathbb{R}^m$ is a measurable set, we denote by $\mathcal{L}(E)$ the family of all measurable functions f on E for which the *finite* $\int_E f\,d\lambda_m$ exists.

4.7 Proposition. If $A \in \mathcal{A}$, then $\mathcal{L}(A) \subset \mathcal{I}(A)$ and $I(f, A) = \int_A f\,d\lambda_m$ for each $f \in \mathcal{L}(A)$.

Proof. Let $f \in \mathcal{L}(A)$, and set $F(B) = \int_B f\,d\lambda_m$ for each $B \in \mathcal{A}(A)$. According to Example 3.3, F is an amiable function in A, and we show that F is the indefinite integral of f in A. Given $\varepsilon > 0$, there are *extended* real-valued functions g and h on A which are, respectively, upper and lower semicontinuous, and such that $g \leq f \leq h$ and $\int_A (h-g)\,d\lambda_m < \varepsilon/2$ (see[Ru, Theorem 2.25, p.56]). For every $B \in \mathcal{A}(A)$, set

$$M(B) = \frac{\varepsilon|B|}{2(1+|A|)} + \int_B (h-g)\,d\lambda_m,$$

and fix an $x \in A$. There is a $\delta > 0$ such that

$$g(y) < f(x) + \frac{\varepsilon}{2(1+|A|)} \text{ and } h(y) > f(x) - \frac{\varepsilon}{2(1+|A|)}$$

for each $y \in A$ with $|x - y| < \delta$. Thus if $B \in \mathcal{A}(A)$, $x \in B^-$, and $d(B) < \delta$, then

$$\int_B g\,d\lambda_m - \frac{\varepsilon|B|}{2(1+|A|)} < f(x)|B| < \int_B h\,d\lambda_m + \frac{\varepsilon|B|}{2(1+|A|)}.$$

We also have

$$\int_B g\,d\lambda_m \leq F(B) \leq \int_B h\,d\lambda_m,$$

and consequently

$$|f(x)|B| - F(B)| < M(B).$$

As $M(A) < \varepsilon$, we conclude that M is an ε-majorant of the pair (f, F) in A.

A set $C = \Pi_{i=1}^{m}[a_i, a_i + h]$, where $a_i, \dots, a_m$, and $h > 0$ are real numbers, is called a ***closed cube.*** Let $A \in \mathcal{A}$, $x \in A^\circ$, and let F be a function on $\mathcal{A}(A)$. We say that F is ***derivable*** at x if a finite $\lim[F(C_n)/|C_n|]$ exists for each sequence $\{C_n\}$ of closed subcubes of A such that $x \in C_n$ for $n = 1, 2, \dots$, and $\lim d(C_n) = 0$. If all these limits exist, they have the same value, denoted by $F'(x)$.

4.8 Lemma. Let $A \in \mathcal{A}$, $f \in \mathcal{I}(A)$, and let $F = I(f, \cdot)$. Then for almost all $x \in A^\circ$ the function F is derivable at x and $F'(x) = f(x)$.

Proof. Let E be the set of all $x \in A^\circ$ for which either F is not derivable at x, or $F'(x) \neq f(x)$. Then given $x \in E$, we can find a $\beta(x) > 0$ so that for each $\delta > 0$ there is a closed cube $C \subset A$ with x$\in C$, $d(C) < \delta$, and

$$\left|\frac{F(C)}{|C|} - f(x)\right| \geq \beta(x)$$

Fix an integer $n \geq 1$, and let $E_n = \{x \in E : \beta(x) \geq 1/n\}$. In view of Remark 4.2, (ii), there is a thin set $T \subset A^-$ such that given $\varepsilon \in (0, 1/2m)$, we can find an (ε/n)-majorant M of the pair (f, F) in $A - T$. It follows that for each $x \in A - T$ there is a $\delta(x) > 0$ so that

$$|f(x)|C| - F(C)| \leq M(C)$$

for each closed cube $C \subset A$ with $x \in C$ and $d(C) < \delta(x)$; for $r(C) > 1/\varepsilon$. Let $\mathcal{C}$ be the family of all closed cubes $C \subset A$ such that $d(C) < \delta(x_C)$ for some $x_C \in C - T$, and

$$|F(C) - f(x_C)|C|| \geq \frac{|C|}{n}.$$

It is easy to see that $\mathcal{C}$ covers $E_n - T$ in the sense of Vitali. By [Sa, Chapter IV, Theorem 3.1, p.112], there are disjoint cubes $C_1, C_2, \dots$ in $\mathcal{C}$ such that $|(E_n - T) - \cup_{i=1}^{\infty} C_i| = 0$. Since

$$\sum_{i=1}^{p}|C_i| \leq n\sum_{i=1}^{p}|F(C_i) - f(x_{C_i})|C_i|| \leq n\sum_{i=1}^{p} M(C_i)$$

$$= nM(\bigcup_{i=1}^{p} C_i) \leq nM(A) < \varepsilon$$

for each $p = 1, 2, \dots$, we have

$$|E_n - T| \leq |\bigcup_{i=1}^{\infty} C_i| = \sum_{i=1}^{\infty}|C_i| \leq \varepsilon.$$

As $|T| = 0$, the arbitrariness of ε implies that $|E_n| = 0$, and as $E = \cup_{n=1}^{\infty} E_n$, also $|E| = 0$.

4.9 Corollary. If $A \in \mathcal{A}$, then each $f \in \mathcal{I}(A)$ is measurable.

Indeed, since $|A^\cdot| = 0$, the corollary follows from Lemma 4.8 by the standard argument (see, e.g.,[Sa, Chapter IV, Theorem (4.2), p.112]).

The next proposition and its corollaries have proofs completely analogous to those of [P1, Proposition 4.4 and Corollaries 4.5, 4.6].

4.10 Proposition. Let f be a function defined on an admissible set A. Then f belongs to $\mathcal{L}(A)$ if and only if both f and $|f|$ belong to $\mathcal{I}(A)$.

4.11 Corollary. Let f be a function defined on an admissible set A. Then $f = 0$ almost everywhere if and only if $f \in \mathcal{I}(A)$ and $I(f, \cdot) = 0$.

4.12 Corollary. Let g, h, and f_n, $n = 1, 2, \ldots$, be integrable functions in an admissible set A, and let $f = \lim f_n$. Suppose that either of the following conditions holds:

1. $f_n \leq f_{n+1}$, $n = 1, 2, \ldots$, and $\lim I(f_n, A) < +\infty$;

2. $g \leq f_n \leq h$, $n = 1, 2, \ldots$.

Then $f \in \mathcal{I}(A)$ and $I(f, A) = \lim I(f_n, A)$.

If $v = (f_1, \ldots, f_m)$ is a vector field defined in $E \subset \mathbb{R}^m$, we call a ***divergence*** of v ***any*** function g on E such that

$$g(x) = \sum_{i=1}^{m} \frac{\partial f_i(x)}{\partial \xi_i}$$

for each $x \in E^\circ$ at which v is differentiable. Each divergence of a vector field v is denoted by $\operatorname{div} v$.

Note. We use the usual definition of a differentiable map (see, e.g., [Ru, Definition 7.22, p.150]). In particular, differentiable does ***not*** mean continuously differentiable.

4.13 Lemma. Let v be a bounded $\mathcal{H}$-measurable vector field in an open set $U \subset \mathbb{R}^m$ which is differentiable at $x \in U$. Then given $\varepsilon > 0$, there is a $\delta > 0$ such that

$$\left|\operatorname{div} v(x)|B| - \int_{B^\cdot} v \cdot n_B \, d\mathcal{H}\right| < \varepsilon|B|$$

for each admissible set B with $B^- \subset U$, $x \in B^-$, $d(B) < \delta$, and $r(B) > \varepsilon$.

This lemma was proved in [P1, Lemma 5.5], and we quote it here for completeness.

4.14 Theorem. Let A be an admissible set, and let S and T be, respectively, a slight and thin subset of A^-. Let v be a bounded vector field in A^- which is continuous in $A^- - S$ and differentiable in $A^\circ - T$. Then $\operatorname{div} v$ is integrable in A and

$$I(\operatorname{div} v, A) = \int_{A^\cdot} v \cdot n_A \, d\mathcal{H}.$$

Proof. Clearly, v is $\mathcal{H}$-measurable, and so for each $B \in \mathcal{A}(A)$, we can set $F(B) = \int_{B^\cdot} v \cdot n_B \, d\mathcal{H}$. By Example 3.4, F is an amiable function in A, and we show that it is an indefinite integral of $f = \operatorname{div} v$ in A. To this end, choose an $\varepsilon > 0$, and let $M(B) = \varepsilon|B|/(1 + |A|)$ for each $B \in \mathcal{A}(A)$. Clearly, M is a nonnegative additive function in A, and $M(A) < \varepsilon$. If $x \in A^\circ - T$, then by Lemma 4.13, there is a $\delta > 0$ such that

$$|f(x)|B| - F(B)| < \frac{\varepsilon}{1+|A|}|B| = M(B)$$

for each $B \in \mathcal{A}(A)$ with $x \in B^-$, $d(B) < \delta$, and

$$r(B) > \varepsilon > \frac{\varepsilon}{1+|A|}.$$

Thus M is an ε-majorant of the pair (f, F) in $A^\circ - T$, and since $A^\cdot \cup T$ is a thin subset of A^-, the theorem is proved.

Let $E \subset \mathbb{R}^m$ and $\Phi : E \to \mathbb{R}^m$. We say that Φ is a ***regular*** map of E if it can be extended to a C^1-diffeomorphism (also denoted by Φ) of an open neighborhood of E^-. For a regular map Φ, we denote by $\det \Phi$ the determinant of its Jacobi matrix. If Φ is regular, then Φ and $\det \Phi$ are defined uniquely in E^- and $(E^\circ)^-$, respectively, and they both extend continuously to a neighborhood of E^-.

4.15 Lemma. Let Φ be a regular map of $E \subset \mathbb{R}^m$. If E is a slight, thin, or admissible set, then so is $\Phi[E]$, respectively.

Proof. This follows immediately from [Ro, Theorem 29, p.53] and the equality $(\Phi[E])^\cdot = \Phi[E^\cdot]$.

4.16 Theorem. Let $A \in \mathcal{A}$, and let $\Phi : A \to \mathbb{R}^m$ be a regular map. If $f \in \mathcal{I}(\Phi[A])$, then $f \circ \Phi \cdot |\det \Phi|$ belongs to $\mathcal{I}(A)$ and

$$I(f \circ \Phi \cdot |\det \Phi|) = I(f, \Phi[A]).$$

With a few adjustments, the idea of the proof is similar to that of [P1, Theorem 6.3].

Let f be an integrable function in an admissible set A. We say that $x \in A^-$ is a ***singular point*** of f if there is a disjoint sequence $\{B_n\}$ in $\mathcal{A}(A)$ such that $\inf r(B_n \cup \{x\}) > 0$, $\lim d(B_n \cup \{x\}) = 0$, and

$$\sum_{n=1}^{\infty} |I(f, B_n)| = +\infty.$$

The set of all singular points of f is denoted by $S(f)$.

4.17 Proposition. Let f be an integrable function in an admissible set A. Then the set $S(f)$ of the singular points of f is thin.

Proof. Proceeding towards a contradiction, assume that $S(f)$ is not thin. For $k = 1, 2, \ldots$, let S_k be the set of all points $x \in A^-$ for which there is a disjoint sequence $\{B_n\}$ in $\mathcal{A}(A)$ such that $\inf r(B_n \cup \{x\}) > 1/k$, $\lim d(B_n \cup \{x\}) = 0$, and

$$\sum_{n=1}^{\infty} |I(f, B_n)| = +\infty.$$

Since $S(f) = \cup_{k=1}^{\infty} S_k$, we see that S_k is not thin for some integer $k \geq 1$. Let $F = I(f, \cdot)$, and find a thin set $T \subset A^-$ and a $(1/k)$-majorant M of the pair (f, F) in $A - T$. As S_k is not thin, there is an $x \in S_k - T$. Let $\{B_n\}$ be a disjoint sequence in $\mathcal{A}(A)$ associated with the point x. By our assumption, the point x is a cluster point of $A - T$. Thus $A - T$ contains disjoint countable sets C_n with $x \in C_n^-$, $n = 1, 2, \ldots$, and we let $D_n = (B_n \cup C_n)$. It follows directly from Definition 4.1 that $F(D_n) = F(B_n)$; for countable sets are thin. Moreover, by making the C_n's sufficiently small, we may assume that $d(D_n) = d(B_n \cup \{x\})$, and hence also $r(D_n) = r(B_n \cup \{x\})$. Consequently, there is an integer $N \geq 1$ such that

$$|f(x)|D_n| - F(D_n)| \leq M(D_n)$$

for each $n \geq N$. From this we obtain that

$$\sum_{n=N}^{p} |F(B_n)| = \sum_{n=N}^{p} |F(D_n)| \leq |f(x)| \sum_{n=N}^{p} |D_n| + \sum_{n=N}^{p} M(D_n) =$$

$$|f(x)| \cdot |\bigcup_{n=N}^{p} D_n| + M(\bigcup_{n=N}^{p} D_n)$$

$$\leq |f(x)| \cdot |A| + M(A) < |f(x)| \cdot |A| + 1/k$$

for all $p \geq N$, This is a contradiction, for $\sum_{n=1}^{\infty} |F(B_n)| = +\infty$.

REFERENCES

B A.S. Besicovitch, *On sufficient conditions for a function to be analytic, and behaviour of analytic functions in a neighbourhood of non-isolated singular points*, Proc. London Math. Soc., 22(1931), 1-9.

Fa K.J. Falconer, *The geometry of fractal sets,* Cambridge Univ. Press, Cambridge 1985.

Fe H. Federer, *Geometric measure theory,* Springer-Verlag, New York, 1969.

G E. Giusti, *Minimal surfaces and functions of bounded variation,* Birkhauser, Basel, 1984.

Ha P.R. Halmos, *Measure theory,* Van Nostrand, New York, 1950.

He R. Henstock, *Lectures on the theory of integration,* World Scientific, Singapore, 1988.

JK1 J. Jarnik and J. Kurzweil, *A nonabsolutely convergent integral which admits transformation and can be used for integration on manifolds,* Czechoslovak Math. J., 35(1985), 116-139.

JK2 J. Jarnik and J. Kurzweil, ***A new and more powerful concept of the PU-integral,*** Czechoslovak Math. J., 38(1988), 8-48.

M J. Mařik, ***The surface integral,*** Czechoslovak Math. J., 6(1956), 522-558.

MM J. Mařik and J, Matyska, ***On a generalization of the Lebesgue integral in*** E^m, Czechoslovak Math. J., 15(1965), 261-269.

P1 W.F.Pfeffer, ***The multidimensional fundamental theorem of calculus.*** J. Australian Math. Soc., 43(1987), 143-170.

P2 W.F. Pfeffer, ***Stokes theorem for forms with singularities,*** C. R. Acad. Sci. Paris, Sér. I, 306(1988), 589-592.

PY W.F. Pfeffer and Wei-Chi Yang, ***A multidimensional variational integral and its extensions,*** Proceedings of Twelfth Summer Symposium in Real Analysis August 9-12 1988.

Ro C.A. Rogers, ***Hausdorff measures,*** Cambridge Univ. Press, Cambridge, 1970.

Ru W. Rudin, ***Real and complex analysis,*** McGraw-Hill, New York, 1987.

Sa S. Saks, ***Theory of the integral,*** Dover, New York, 1964.

Sh V.L. Shapiro, ***The divergence theorem for discontinuous vector fields,*** Ann. Math., 68(1958), 604-624.

University of California
Davis, California 95616, USA

Some properties of dyadic primitives

V.A.Skvortsov

We consider here some properties of integrals which integrate finite dyadic derivatives and also some other properties of functions determined by the behaviour of the extreme dyadic derivates.

The dyadic (or binary) derivative is defined as the derivative with respect to a sequence of dyadic nets $\{\mathcal{N}_n\} = \mathcal{N}$, the set $\mathcal{N}_n (n = 0, 1, 2, \ldots)$ consisting of dyadic intervals

$$\delta_j^{(n)} = (\frac{i}{2^n}, \frac{i+1}{2^n}), \; i = 0, 1, \ldots, 2^n - 1,$$

which we call the dyadic intervals of order n.

For any dyadic irrational x, there exists a unique sequence of dyadic intervals $\{\delta_{j(x)}^{(n)}\}_{n=0}^{\infty}$ convergent to x. For a dyadic rational x, there exist two decreasing sequences of dyadic intervals for which x is a common end point, starting from some n.

Given an interval $\delta = (a, b)$, we shall use the notation

$$\Delta F(\delta) = F(b) - F(a).$$

The dyadic derivative in a dyadic irrational x of a function $F(x)$ on $[0, 1]$ is defined as

$$D_{\mathcal{N}} F(x) = \lim_{n \to \infty} \frac{\Delta F(\delta_{j(x)}^{(n)})}{|\delta_{j(x)}^{(n)}|}$$

where $\{\delta_{j(x)}^{(n)}\}$ is the sequence convergent to x. In the case of a dyadic rational x the dyadic derivative can be defined as the common limit of the same ratio for both decreasing sequences of intervals related to x. However in this paper we can in many cases avoid considering the dyadic derivative in dyadic rationals, because the behaviour of the derivative on a countable set is irrelevant in those cases.

The lower derivate $\underline{D}_{\mathcal{N}} F(x)$ and upper derivate $\overline{D}_{\mathcal{N}} F(x)$ can be defined as the upper and lower limits of the above ratio.

F is said to be continuous on $[0, 1]$ with respect to the sequence $\mathcal{N}$ of dyadic sets ($\mathcal{N}$-continuous) if $\Delta F(\delta_{j(x)}^{(n)}) \to 0$ as $n \to \infty$. for any decreasing sequence of dyadic intervals $\{\delta_{j(x)}^{(n)}\}_{n=1}^{\infty}$. Here it is essential to consider all the decreasing sequences of intervals, including those having a common rational end-point.

There are several definitions of integrals wide enough to integrate any everywhere finite dyadic derivative (or one which is finite everywhere except on a countable set of points). The particular interest of the integration of these derivatives comes from the fact that this problem is closely related to another one, namely, the problem of recovering coefficients of an everywhere (or everywhere except on a countable set) convergent Walsh or Haar series from its sum by Fourier formulas.

This relation is based on the observation that the partial sums of order 2^n of either a Walsh or a Haar series can be expressed in the form

$$S_{2^n}(x) = \frac{\Delta F(\delta^{(n)}_{j(x)})}{|\delta^{(n)}_{j(x)}|}.$$

Here $\{\delta^{(n)}_{j(x)}\}$ is a sequence of dyadic intervals converging to x (—we can restrict ourselves to considering only dyadic irrational x) and $F(x)$ is the sum of the (term by term) integrated series which is always convergent—at least on the dyadic rationals. So the problem of calculating the coefficients from the sum of a Walsh or Haar series is practically the same as that of recovering the primitive from a dyadic derivative [1].

A simple way of solving the problem of integrating the dyadic derivative is to introduce a Perron-type integral (see [2]).

A function M from $[0,1]$ to $I\!R$ is said to be a $P_{\mathcal{N}}$-major function for a function f on $[0,1]$ if

- $M(0) = 0$
- M is $\mathcal{N}$-continuous on $[0,1]$,
- $\underline{D}_{\mathcal{N}}M(x) \geq f(x)$ and $\underline{D}_{\mathcal{N}}M(x) > -\infty$ everywhere except on a countable subset of $[0,1]$.

A function m from $[0,1]$ to $I\!R$ is a $P_{\mathcal{N}}$-minor function for f on $[0,1]$ if and only if $-m$ is a $P_{\mathcal{N}}$-major function for $-f$ on $[0,1]$.

It can be proved [3] that if M and m are any pair of major and minor functions for f then $M - m$ is increasing and so $M(1) \geq m(1)$. If at least one of such a pair for f exists and if $\inf M(1) = \sup m(1) = I$ where the inf and sup are taken over the sets of major and minor functions for f respectively, then f is said to be $P_{\mathcal{N}}$-integrable (dyadic Perron integrable); and I is called the $P_{\mathcal{N}}$ integral of f on $[0,1]$, written $P_{\mathcal{N}} \int_0^1 f$.

It is clear that if f is $P_{\mathcal{N}}$-integrable then the indefinite $P_{\mathcal{N}}$-integral (or $P_{\mathcal{N}}$-primitive) of f is also defined. We shall denote it by $F(x) = P_{\mathcal{N}} \int_0^x f$.

A Henstock-Kurzweil type integral integrating dyadic derivatives has also been introduced [4]. Call this the $H_{\mathcal{N}}$-integral. It can be considered now as a special case of a general Henstock integral with respect to an abstract derivative base [5,6].

One can get as a special case of a general situation [6] that the $H_{\mathcal{N}}$-integral is equivalent to the corresponding $P_{\mathcal{N}}$-integral. It can also be shown that the equivalence in this particular case of a base does not depend on whether we use only $\mathcal{N}$-continuous major and minor functions in the definition of the $P_{\mathcal{N}}$-integral or arbitrary majorants and minorants.

In the case of ordinary Perron integration we have two more equivalent definitions of this integral, namely descriptive and constructive definitions for special (or restricted or simple) Denjoy totalisation ($\mathcal{D}^*$-integral) [7]. The question arises whether one can get an equivalent Denjoy type definition for the $P_{\mathcal{N}}$-integral (and $H_{\mathcal{N}}$-integral). We recall first some definitions used in the Denjoy descriptive approach (see [7]).

A function F is said to be absolutely continuous on a set E, or, simply, AC on E, if, given any $\varepsilon > 0$, there exists $\eta > 0$ such that, for every sequence of non-overlapping intervals $\{(a_k, b_k)\}$ whose end-points belong to E, the inequality $\sum_k(b_k - a_k) < \eta$ implies $\sum_k |\Delta F((a_k, b_k))| < \varepsilon$.

A function F is said to be generalised absolutely continuous on E, or ACG on E, if F is continuous on E and if E is the union of a sequence of sets $\{E_n\}$ on each of which F is AC. A function F is said to possess the Lusin N-property on a set E if $|F(H)| = 0$ for every set $H \subset E$, $|H| = 0$. It is known [7] that a function which is ACG on a set necessarily possesses the Lusin N-property.

A function f is said to be $\mathcal{D}$-integrable (integrable in the general or wide Denjoy sense) on (a, b) if there exists a function F which is ACG on (a, b) and which has f for its approximate derivative (see [7]) almost everywhere. The function F is then called the indefinite $\mathcal{D}$-integral of f on (a, b). Its increment $\Delta F((a, b))$ is termed the definite $\mathcal{D}$-integral.

If we replace in this definition the approximate derivative by the ordinary one we get a definition of the Khintchine integral [8]. If we replace in the last definition the class ACG by the more narrow class ACG$_*$ we get a definition of the $\mathcal{D}^*$-integral (special or restricted Denjoy integral). (For the definition of ACG$_*$, see [7].)

A function F is said to be of bounded variation on E, or VB on E, if

$$V(F; E) = \sup \sum_k |\Delta F((a_k, b_k))| < \infty$$

where the sup is taken over all the sequences of non-overlapping intervals whose end-points belong to E. $V(F; E)$ is termed the weak variation of F on E. A function F is said to be of generalised bounded variation on E, or VBG on E, if E is the union of a sequence of sets $\{E_n\}$ on each of which F is VB.

The possibility of getting a Denjoy-type descriptive definition for the $P_{\mathcal{N}}$-integral depends on some properties of the $P_{\mathcal{N}}$-primitive and the $P_{\mathcal{N}}$-major and minor functions. The ordinary Perron major and minor functions are known to be VBG [7]. It is also known that the ordinary P-primitive is ACG$_*$ and so possesses the Lusin N-property.

Coming back to dyadic derivatives and restricting ourselves to the task of integrating just the everywhere finite dyadic derivative, we do have a suitable definition of the Denjoy-type integrals. A descriptive definition of such an integral in terms of the generalised ACG class was given by the author in [9]. An equivalent constructive definition of the same integral using transfinite induction was introduced by the author in [10]. In fact the same transfinite sequences of operations (but in terms of martingales) is used by J.-P. Kahane in [11] to integrate any exact finite dyadic derivative.

The principal operations involved in this construction of the Denjoy-type totalisation are as follows:

- Lebesgue integration on a set;
- if f is summable on a closed set $E \subset [a,b]$ and the Denjoy-type integral has already been defined on every contiguous interval (α_n, β_n), the series $\sum_n \int_{\alpha_n}^{\beta_n} f$ being convergent, then this Denjoy-type integral is defined on $[a,b]$ as
$$\int_a^b f = \mathcal{L}\int_E f + \sum_n \int_{\alpha_n}^{\beta_n} f.$$

It is clear that these operations give the indefinite integral possessing the Lusin N-property. The same is true for the indefinite integral defined descriptively in terms of ACG functions. But it is not the case with the $P_{\mathcal{N}}$- (and $H_{\mathcal{N}}$-) primitive.

We shall see later on that the indefinite $P_{\mathcal{N}}$-integral can fail to possess the Lusin N-property and so can fail to belong to any of the ACG-type classes. Moreover, $P_{\mathcal{N}}$-major functions can fail to be VBG, in sharp contrast to the ordinary case. We shall get all these and some other facts as consequences of one construction of a function which is a $P_{\mathcal{N}}$-primitive and also a $P_{\mathcal{N}}$-majorant having some peculiar features. To construct such a function is the main task of this paper.

We shall start with the following auxiliary function to be defined for each $\delta_i^{(k)}$. ($\bar{\delta}$ denotes closure of δ.)

(1)
$$\begin{aligned} \phi_i^{(k)}(x) &= 0 \text{ if } x = \frac{i}{2^k},\ x = \frac{i+1}{2^k} \text{ or } x \notin \delta_i^{(k)}, \\ &= 1 \text{ if } x \in \delta_{4i+1}^{(k+2)}, \\ \text{and} &\quad \text{linear on } \bar{\delta}_{4i}^{(k+2)} \text{ and } \bar{\delta}_{2i+1}^{(k+1)}. \end{aligned}$$

Observe that

(2)
$$\begin{aligned} \Delta\phi_i^{(k)}(\delta_i^{(k)}) &= \Delta\phi_i^{(k)}(\delta_{4i+1}^{(k+2)}) = 0, \\ \Delta\phi_i^{(k)}(\delta_{2i}^{(k+1)}) &= \Delta\phi_i^{(k)})(\delta_{4i}^{(k+2)}) = 1, \\ \Delta\phi_i^{(k)}(\delta_{2i+1}^{(k+1)}) &= -1. \end{aligned}$$

Now put

$$\begin{aligned}
\Phi_0(x) &= \textstyle\sum_{i=0}^{3} \phi_i^{(2)}(x), \\
\Phi_1(x) &= \tfrac{1}{4}\textstyle\sum_{i=0}^{3}(\phi_i^{(2)}(x) + \phi_{4i+1}^{(4)}(x) + \phi_{4^2 i+4+1}^{(6)}(x) + \phi_{4^3 i+4^2+4+1}^{(8)}(x)), \\
&\vdots \\
\Phi_n(x) &= 4^{-n}\textstyle\sum_{i=0}^{3}\sum_{k=0}^{4^n-1} \phi_{4^k i+\sum_{r=0}^{k-1} 4^r}^{(2k+2)}(x).
\end{aligned}$$

(Here, $\sum_{r=0}^{-1} = 0$.) We consider $\Phi_n(x)$ to be defined on the real line having support on $(0,1)$ and vanishing outside $(0,1)$. The construction of $\Phi_{(n)}(x)$ for $n = 1$ is illustrated in fig.1.

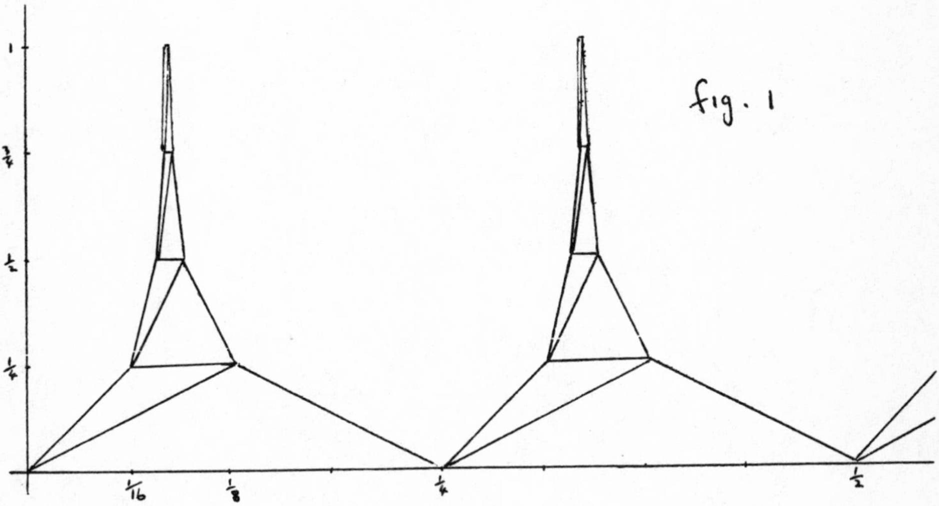

fig. 1

It follows from (1) and (2) that

(3) $$|\Phi_n(x)| \leq 1 \text{ everywhere,}$$

(4) $$\Phi_n(x) = 1 \text{ if } x \in A_n \equiv \bigcup_{i=0}^{3} \delta_{4^{4^n} i+\sum_{r=0}^{4^n-1} 4^r}^{(2.4^n+2)},$$

(5) $$0 \leq \Delta\Phi_n(\delta_{j(x)}^{(k)}) \leq 4^{-n} \text{ for all } \delta_{j(x)}^{(k)}) \ni x, \text{ if } x \in A_n.$$

Notice that

(6) $$|A_n| \leq \frac{1}{4} \text{ for any } n \geq 1.$$

For some dyadic intervals $\delta = (a,b)$ we shall use the following linear mapping $\delta \to (0,1)$:

(7) $$\begin{aligned}
l(x,\delta) &= |\delta|^{-1}(x-a) \text{ if } s(\delta) = 1, \\
&= |\delta|^{-1}(b-x) \text{ if } s(\delta) = -1,
\end{aligned}$$

where the sign function $s(\delta)$ is defined below.

We define by induction a double sequence of dyadic intervals

$$\{\gamma_j^{(n)}\},\ j=0,1,\ldots,4^n-1,\ n=1,2,\ldots.$$

Put

$$\gamma_j^{(1)}=\delta_{4j+1}^{(4)},\ j=0,1,2,3.$$

Now suppose $\gamma_j^{(n)}$ is defined for some $n\geq 1,\ 0\leq j\leq 4^n-1$. Then we define

$$\gamma_{4j+i}^{(n+1)},\ i=0,1,2,3$$

to be the inverse image of

$$\delta^{(2.4^n+2)}_{4^{4^n}i+\sum_{r=0}^{4^n-1}4^r}$$

under $l(x,\gamma_j^{(n)})$. (See (7).) Thus we define $\gamma_j^{(n+1)}$ for all $j=0,1,\ldots,4^{n+1}-1$.

Obviously, for fixed n the intervals $\gamma_j^{(n)}, 0\leq j\leq 4^n-1$, are disjoint. Put

(8)
$$Q_n=\bigcup_{j=0}^{4^n-1}\gamma_j^{(n)},\ n=1,2,\ldots.$$

Since, by (6),

$$\sum_{i=0}^{3}|\gamma_{4j+i}^{(n+1)}|\leq\frac{1}{4}|\gamma_j^{(n)}|$$

then

(9)
$$|Q_{n+1}\leq\frac{1}{4}|Q_n|$$

It can easily be deduced from (3), (4), (5), (7) and the above definition of $\gamma_j^{(n)}$ that the functions

$$s(\gamma_j^{(n)})\Phi_{(n)}(l(x,\gamma_j^{(n)})$$

(i.e. $\Phi_n(|\gamma_j^{(n)}|^{-1}(x-a_j^{(n)}))$ and $-\Phi_n(|\gamma_j^{(n)}|^{-1}(b_j^{(n)}-x))$, where $(a_j^{(n)},b_j^{(n)})=\gamma_j^{(n)})$ have the following properties:

(10)
$$|s(\gamma_j^{(n)})\Phi_n(l(x,\gamma_j^{(n)}))|\leq 1 \text{ everywhere;}$$

(11)
$$s(\gamma_j^{(n)}\Phi_n(l(x,\gamma_j^{(n)}))=0 \text{ if } x\notin\gamma_j^{(n)};$$

(12)
$$s(\gamma_j^{(n)})\Phi_n(l(x,\gamma_j^{(n)}))=s(\gamma_j^{(n)}) \text{ if } x\in Q_{n+1};$$

(13)
$$0\leq\Delta s(\gamma_j^{(n)})\Phi_n(l(\delta,\gamma_j^{(n)}))\leq\frac{1}{4^n}$$

if $x\in Q_{n+1}$ and δ is any dyadic interval containing x;

(14)
$$\Delta s(\gamma_j^{(n)})\Phi_n(l(\delta,\gamma_j^{(n)}))=0 \text{ if } \gamma_j^{(n)}\subset\delta.$$

Now, for $x \in [0,1]$ let

$$F_n(x) = \frac{1}{2}\Phi_0(x) + \sum_{k=1}^{n}\sum_{j=0}^{4^k-1} 2^{-k} s(\gamma_j^{(k)})\Phi_k(l(x,\, \gamma_j^{(k)})) \tag{15}$$

and define $s(\gamma)$ by induction:

$$s(\gamma_i^{(1)}) = (-1)^{(i+1)},\ i = 0,1,2,3; \tag{16}$$

$$\begin{aligned} s(\gamma_{4j+i}^{(n+1)}) &= (-1)^i s(\gamma_j^{(n)}) \text{ if } j = 4p \text{ or if } j = 4p+2 \text{ and } s(\gamma_j^{(n)}) = s(\gamma_{j-1}^{(n)}), \\ &= (-1)^{i+1} s(\gamma_j^{(n)}) \text{ if } j = 4p+3 \text{ or if } j = 4p+1 \text{ and } s(\gamma_j^{(n)}) = s(\gamma_{j+1}^{(n)}), \\ &= (-1)^{i+1+[i/2]} s(\gamma_j^{(n)}) \text{ if } j = 4p+1 \text{ and } s(\gamma_j^{(n)}) = -s(\gamma_{j+1}^{(n)}), \\ &= (-1)^{i+[i/2]} s(\gamma_j^{(n)}) \text{ if } j = 4p+2 \text{ and } s(\gamma_j^{(n)}) = -s(\gamma_{j-1}^{(n)}). \end{aligned}$$

Here, $0 \le j \le 4^n - 1$; $i = 0,1,2,3$; and $[i/2]$ is the integer part of $i/2$.

fig. 2

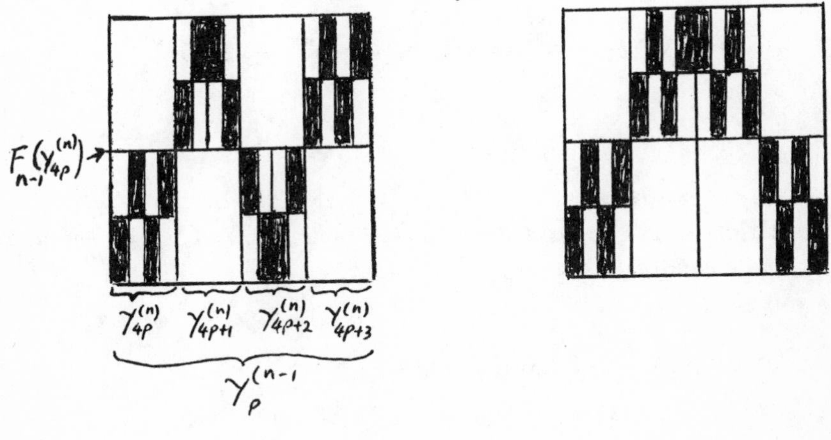

The construction of F_n and the definition (16) are illustrated schematically by the diagrams in fig.2, where two of four possible cases are shown. The others are symmetrical to those shown with respect to the mid-lines of the big rectangles. In the diagram each marked rectangle shows where the graph of F_{n+1} lies relative to that of F_n, the latter being symbolised by the mid-line of the parent rectangle. Similar diagrams were used in [12] where they were attributed to J.Foran.

Observe that, by (16), $s(\gamma_{4j}^{(n+1)}) = s(\gamma_j^{(n)})$ if $j = 4p$ and hence

$$s(\gamma_{4^k j}^{(n+k)}) = s(\gamma_{4j}^{(n)}) \text{ for any } j \ge 0,\ k \ge 0. \tag{17}$$

Also, by (16), $s(\gamma_{4j+3}^{(n+1)}) = s(\gamma_j^{(n)})$ if $j = 4p+3 = 4(p+1)-1$, and hence

(18) $$s(\gamma_{4^{k+1}j-1}^{(n+k)}) = s(\gamma_{4j-1}^{(n)}) \text{ for any } j \geq 1,\ k \geq 0.$$

Put

(19) $$F(x) = \lim_{n\to\infty} F_n(x) = \frac{1}{2}\Phi_0(x) + \sum_{k=1}^{\infty}\sum_{j=0}^{4^k-1} 2^{-k} s(\gamma_j^{(k)})\Phi_k(l(x,\gamma_j^{(k)}))$$

and

(20) $$Q = \bigcap_{n=1}^{\infty} Q_n.$$

As the intervals $\{\gamma_j^{(k)}\}_j$ are disjoint, it follows from (10) that the series in (19) converges uniformly and so $F(x)$ is continuous. Q is obviously a perfect set and, by (9), $|Q| = 0$.

Note that the sum in (19) is in fact finite for each $x \notin Q$. So $F(x)$ is a countably piecewise linear function on each interval contiguous to Q. Hence $F'(x) = D_{\mathcal{N}}F(x)$ exists everywhere outside Q except on some dyadic rationals.

Now consider $x \in Q$. Then for any dyadic interval δ, $x \in \delta$, we choose n so that $\gamma_i^{(n+1)} \subset \delta \subset \gamma_j^{(n)}$. For such δ and n we get by (19), (11)—(14),

(21) $$0 \leq \Delta F(\delta) = 2^{-n}\Delta s(\gamma_i^{(n)})\Phi_n(l(\delta,\gamma_j^{(n)})) \leq 2^{-n}4^{-n}.$$

This inequality implies

(22) $$D_{\mathcal{N}}F(x) \geq 0 \text{ if } x \in Q.$$

Therefore we can state

Proposition 1. *If $F(x)$ is defined by (19) then $\underline{D}_{\mathcal{N}}F(x) > -\infty$ everywhere except on some countable set.*

Now put

(23) $$\begin{aligned} f(x) &= F'(x) \text{ if } x \notin Q \text{ and } F'(x) \text{ exists },\\ &= 0 \text{ if } x \in Q \text{ or } F'(x) \text{ does not exist.}\end{aligned}$$

Then we get that $\underline{D}_{\mathcal{N}}F(x) \geq f(x)$ everywhere except on some countable set. Since $F(x)$ is also continuous and $F(0) = 0$, we have

Proposition 2. *$F(x)$ is a $P_{\mathcal{N}}$-major function for $f(x)$.*

We now consider some other properties of F. Notice that

$$F_n(x) = F_{n-1}(x) + \sum_{j=0}^{4^n-1} 2^{-n}s(\gamma_j^{(n)})\Phi_n(l(x,\gamma_j^{(n)})).$$

Hence $F_n(x)$ is a constant on each $\gamma_{4j+i}^{(n+1)}$. Denote this constant by $F_n(\gamma_{4j+1}^{(n+1)})$. Moreover, this constant is the same for $i = 0,1,2,3$ if j is fixed, and

(24) $$F_n(\gamma_{4j+i}^{(n+1)}) = F_{n-1}(\gamma_j^{(n)}) + 2^{-n}s(\gamma_j^{(n)})\Phi_n(l(x,\gamma_j^{(n)})).$$

Now fix two neighbouring intervals $\gamma_{j-1}^{(n)}$ and $\gamma_j^{(n)}$ belonging to the same $\gamma_p^{(n-1)}$. This implies that $j = 4p+m$, $m = 1,2,3$. Consider for any $k \geq 1$ the intervals $\gamma_{4^k j-1}^{(n+k)}$ and $\gamma_{4^k j}^{(n+k)}$ which are the furthest right of $\gamma_i^{(n+k)}$ in $\gamma_{j-1}^{(n)}$ and the furthest left of $\gamma_i^{(n+k)}$ in $\gamma_j^{(n)}$ respectively. Again,

$$F_{n-1}(\gamma_{4p+m-1}^{(n)}) = F_{n-1}(\gamma_{4p+m}^{(n)}). \tag{25}$$

If

$$s(\gamma_{j-1}^{(n)}) = -s(\gamma_j^{(n)}) \tag{26}$$

then, considering all possible cases in (16) we can see that, for $\gamma_{4j-1}^{(n+1)}$ ($= \gamma_{4(j-1)+3}^{(n+1)}$) and $\gamma_{4j}^{(n+1)}$,

$$\begin{aligned} s(\gamma_{4j-1}^{(n+1)}) &= -s(\gamma_{j-1}^{(n)}) \text{ and} \\ s(\gamma_{4j}^{(n+1)}) &= -s(\gamma_j^{(n)}). \end{aligned} \tag{27}$$

If

$$s(\gamma_{j-1}^{(n)}) = s(\gamma_j^{(n)}) \tag{28}$$

which, by (16), could occur only if $j = 4p+2$ and $j-1 = 4p+1$, then

$$s(\gamma_{4j-1}^{(n+1)}) = s(\gamma_j^{(n)}) \text{ and } s(\gamma_{4j}^{(n+1)}) = s(\gamma_j^{(n)}). \tag{29}$$

Put

$$\alpha = \bigcap_{k=0}^{\infty} \gamma_{4^k j-1}^{(n+k)}, \quad \beta = \bigcap_{k=0}^{\infty} \gamma_{4^k j}^{(n+k)}. \tag{30}$$

By (20), $\alpha, \beta \in Q$, and, since $(\alpha, \beta) \cap Q$ is empty, then (α, β) is an interval contiguous to Q, $(\alpha, \beta) \subset \gamma_p^{(n-1)}$.

Now in the case of (26) we find from (19), (24), (27), (17), (18), (12) that

$$F(\alpha) = F_{n-1}(\gamma_{j-1}^{(n)}) + s(\gamma_{(j-1)}^{(n)})(\frac{1}{2^n} - \sum_{k=n+1}^{\infty} \frac{1}{2^k}) = F_{n-1}(\gamma_{j-1}^{(n)}),$$

$$F(\beta) = F_{n-1}(\gamma_j^{(n)}) + s(\gamma_j^{(n)})(\frac{1}{2^n} - \sum_{k=n+1}^{\infty} \frac{1}{2^k}) = F_{n-1}(\gamma_j^{(n)})$$

and so, by (25),

$$F(\alpha) = F(\beta) = F_{n-1}(\gamma_j^{(n)}), \tag{31}$$

and this is always the case if $j = 4p+1$ or $j = 4p+3$.

By the same arguments in the case of (28) we have, using (29) instead of (27),

$$\begin{aligned} F(\alpha) = F(\beta) &= F_{n-1}(\gamma_j^{(n)}) + s(\gamma_j^{(n)})(\tfrac{1}{2^n} + \textstyle\sum_{k=n+1}^{\infty} \tfrac{1}{2^k}) \\ &= F_{n-1}(\gamma_j^{(n)}) + s(\gamma_j^{(n)})\tfrac{1}{2^{n-1}}. \end{aligned}$$

It is easy to show that the endpoints of any proper contiguous interval (α, β) can be represented in the form (30) for some $j = 4p + m$, $m = 1, 2, 3$. So we have proved that, for any proper contiguous interval (α, β),

$$F(\alpha) = F(\beta). \tag{32}$$

It is also easy to see that for two improper intervals $(0, \beta)$ and $(\alpha, 1)$ we have

$$F(0) = F(\beta) = 0, \; F(\alpha) = F(1) = 1. \tag{33}$$

We can also observe from (31) that any interval $\gamma_p^{(n)}$ contains a point $x \in Q$ such that $F(x) = F_n(\gamma_{4p}^{(n+1)})$.

Now consider intervals $\gamma_j^{(n)} \subset \gamma_p^{(n-1)}$ for $j = 4p + m$, $m = 0, 1, 2, 3$. As $s(\gamma_j^{(n)})$ changes sign at least twice when j varies from $4p$ to $4p + 3$ (see (16)), then, from (24), (25), (12) and from the above observation it follows that for the weak variation of F we have

$$V(F,\, Q \cap \gamma_p^{(n-1)}) \geq \frac{1}{2^{n-1}}.$$

This holds for any n. Now each $\gamma_j^{(n)}$ contains 4^k intervals $\gamma_i^{(n+k)}$, $4^k j \leq i \leq 4^k(j+1) - 1$ for any $k \geq 1$. Thus

$$V(F,\, Q \cap \gamma_j^{(n)}) \geq \sum_{i=4^k j}^{4^k(j+1)-1} V(F,\, Q \cap \gamma_i^{(n+k)}) \geq 4^k \frac{1}{2^{n+k}} = 2^{k-n}.$$

Consequently $F(x)$ is not VB on any portion of the set Q. It is known (see [7]) that, for any VBG function on E, any closed subset of E contains a portion on which it is VB. So we have proved

Proposition 3. *F is not VBG on Q.*

It is not difficult to change slightly the value of F in some small neighbourhood of each point outside Q where $F'(x)$ does not exist in order to make F differentiable everywhere outside Q, but without affecting the value of F on Q or the value of $\underline{D}_{\mathcal{N}}F(x)$ on Q. Then the inequality in Proposition 1 holds everywhere and, combining it with Proposition 3 we have

Theorem 1. *There exists a continuous function F for which the inequality $\underline{D}_{\mathcal{N}}F(x) > -\infty$ everywhere does not imply that F is VBG.*

Remember [7] that for the ordinary lower derivate and many other generalised lower derivates this implication does hold. Note, on the other hand, that certain theorems which apply to ordinary derivates also hold for dyadic derivates. For example,

- $\overline{D}_{\mathcal{N}}F(x) \geq 0$ on $[0,\, 1]$ implies monotonicity (see [3]),
- $-\infty < \underline{D}_{\mathcal{N}}F(x) \leq \overline{D}_{\mathcal{N}}F(x) < \infty$ on $[0,\, 1]$ implies that F is generalised Lipschitz, i.e. $[0,\, 1] = \bigcup_{i=1}^{\infty} E_i$ and F satisfies a Lipschitz condition on each E_i [9].

(The term "generalised Lipschitz" was introduced in [13].)

Now we define the non-decreasing function

$$R_n(x) = \frac{4}{2^n|Q_{n+1}|}\int_0^x \chi_{Q_{n+1}}(t)dt.$$

(χ_E is the indicator of E.) We get, for any $\gamma_j^{(n+1)}$,

$$\Delta R_n(\gamma_j^{(n+1)}) = \frac{4|\gamma_j^{(n+1)}|}{2^n|Q_{n+1}|} = \frac{4}{2^n}\frac{1}{4^{n+1}} = \frac{1}{2^n 4^n}.$$

For any dyadic interval $\delta \supset \gamma_j^{(n+1)}$ we have

$$\Delta R_n(\delta) \geq \Delta R_n(\gamma_j^{(n+1)}) = \frac{1}{2^n 4^n}. \tag{34}$$

Put

$$S_k(x) = \sum_{n=k}^{\infty} R_n(x) \text{ and } m_k(x) = F(x) - S_k(x).$$

We have $S_k(1) = \sum_{n=k}^{\infty} \frac{1}{2^{n-2}}$ and hence

$$0 \leq S_k(x) \leq \frac{1}{2^{k-3}} \text{ everywhere on } (0,\ 1). \tag{35}$$

Now fix k and let $x \in Q$. There exists $\gamma_j^{(k)} \ni x$. Take any dyadic interval δ such that $x \in \delta \subset \gamma_j^{(k)}$. Choose $n \geq k$ such that $\gamma_j^{(n+1)} \subset \delta \subset \gamma_m^{(n)}$. By (21) and (34) we have

$$\Delta m_k(\delta) = \Delta F(\delta) - \Delta S_k(\delta) \leq \frac{1}{2^n 4^n} - \frac{1}{2^n 4^n} = 0.$$

Hence

$$D_{\mathcal{N}} m_k(x) \leq 0 \text{ if } x \in Q.$$

It is clear that $S_k'(x)$ exists everywhere outside Q except for some dyadic rationals and $S_k'(x) \geq 0$. Then, by (23),

$$D_{\mathcal{N}} m_k(x) = m_k'(x) = F'(x) - S'(x) \leq F'(x) = f(x)$$

if $x \notin Q$ and $m_k'(x)$ exists. Thus we have proved that $m_k(x)$, for any k, is a $P_{\mathcal{N}}$-minor function for $f(x)$.

Proposition 2 and (35) now imply

Proposition 4. *The function f defined by (23) is $P_{\mathcal{N}}$-integrable (and consequently $H_{\mathcal{N}}$-integrable). Moreover,*

$$F(x) = P_{\mathcal{N}} \int_0^x f(t)dt.$$

Now consider the function

$$(36) \qquad F(x,\, Q) = \begin{cases} F(x) \text{ if } x \in Q, \\ \text{and linear in the closure of each interval} \\ \text{contiguous to } Q. \end{cases}$$

By (32) and (33), $F(x,\, Q)$ is constant on each interval contiguous to Q and $F(0,\, Q) = 0$, $F(1,\, Q) = 1$. Remember that $|Q| = 0$. Thus $F(x,\, Q)$ is singular and does not possess the Lusin N-property. As $F(x) = F(x,\, Q)$ for $x \in Q$, we have

Theorem 2. *There exists a $P_{\mathcal{N}}$-integrable (and $H_{\mathcal{N}}$-integrable) function f for which the indefinite integral $F(x)$ fails to possess the Lusin N-property.*

This theorem shows that $F(x)$ cannot be recovered from $f(x)$ by the same kind of constructive operations that are used in the Denjoy totalisation process, as in [10] or [11]. Also, $F(x)$ cannot be ACG. So it is impossible in these terms to get an equivalent Denjoy-type definition for the $P_{\mathcal{N}}$- and $H_{\mathcal{N}}$-integral. Of course, this does not exclude the possibility of getting a constructive definition for the $P_{\mathcal{N}}$-integral, using operations of quite a different type, which can permit the primitive not to have the N-property.

Now put

$$\Psi(x) = F(x) - F(x,\, Q).$$

By (32), (33) and (36),

$$\Psi(x) = 0 \text{ if } x \in Q \text{ and } \Psi(0) = \Psi(1) = 0.$$

$\Psi(x)$ is countable piecewise linear on each interval contiguous to Q. So $\Psi(x)$, being also continuous, is obviously ACG and consequently is an indefinite D-integral of its derivatives $\Psi'(x)$ which is equal to $F'(x) = f(x)$ a.e. Therefore we get

Theorem 3. *There exists a function f on $[0,\, 1]$ both $P_{\mathcal{N}}$-integrable ($H_{\mathcal{N}}$-integrable) and D-integrable (Denjoy integrable in the wide sense) for which*

$$D\int_0^1 f \neq P_{\mathcal{N}} \int_0^1 f.$$

Some of these results were announced in [14].

BIBLIOGRAPHY

1. V.A.Skvortsov, Differentiation with respect to nets and Haar series, Mat. Zametki, 4(1968), 33-44.

2. V.A.Skvortsov, Haar series with convergent subsequences of partial sums, Dokl. Akad. Nauk SSSR, 183(1968), 784-786; Engl. transl., Sov. Math. Dokl. 9(1968), 1969-1971.

3. V.A.Skvortsov, A generalisation of Perron's integral, Vest. Mosk. Univ. Ser.I, 24(1969), 48-51.

4. A.Pacquement, Détermination d'une fonction au moyen de sa dérivée sur un réseau binaire, C.R.Acad.Sci.Paris 284,A(1977), 365-368.

5. B.S.Thomson, Derivation bases on the real line, I, II, Real Analysis Exchange 8(1982/83), 67-207, 278-442.

6. K.M.Ostaszewski, Henstock integration in the plane, Memoirs of AMS 67(1986), N353.

7. S.Saks, Theory of the integral, 2nd English edition, Warsaw 1937.

8. A.Khintchine, Sur une extension de l'intégrale de M. Denjoy, C. R. Acad. Sci. Paris, 162(1916), 287-291.

9. V.A.Skvortsov, Calculation of the coefficients of an everywhere convergent Haar series, Mat. Sb., 117(1968), 349-360; Engl. transl., Math. USSS Sb., 4(1968), 317-327.

10. V.A.Skvortsov, Constructive definition of HD-integral, Moscow Un. Math. Bull. 37(1982) N6, 46-50.

11. J.-P.Kahane, Une theorie de Denjoy des martingales dyadiques, Prepublication, Paris, 1988.

12. A.M.Bruckner, Differentiation of real functions, Lecture Notes in Math., v659, 1978.

13. A.M.Bruckner and Cliff Cordy, Near intersection conditions for path derivative, Real Anal. Exchange 13(1987-88) N1, 16-23.

14. V.A.Skvortsov, Generalised integrals in the theory of trigonometric, Haar and Walsh series, Real Anal. Exchange, 12(1986-87) N1, 59-62.

Moscow State University

Moscow 119899

Analysis of P. Malliavin's proof of non spectral synthesis

J.D. Stegeman

1 Motivation

In 1959, now thirty years ago, Paul Malliavin published his celebrated theorem on the impossibility of spectral synthesis on non compact locally compact abelian groups. His paper [8] is a landmark in the history of spectral synthesis.
Malliavin's paper is written in a concise way and is not easy to read. The proof of his theorem consists of three parts:

I. a group structural argument,

II. a functional analytic argument,

III. a technical construction.

Part III in particular is rather complicated. For instance, in his book [9] (p.126) H. Reiter calls the proof "as difficult as it is profound". And according to E. Hewitt and K.A. Ross ([6], p.602) it is "a short, difficult paper".
Soon after the publication alternative methods of proof were developed; see for instance [7], Chapitre IX, [11], section 7.6, and [10]. Moreover, in 1965 N.Th. Varopoulos found an entirely different approach, using tensor algebras. See, e.g., the expositions in [3] or [6].
No detailed treatment of Malliavin's original proof seems to be readily available in the literature. E.g., in [1] the author "does not provide the details to these constructions" (p.188). Whatever the reason, there can be little doubt that the technicalities of Malliavin's paper are less generally understood than they deserve. We have therefore set out to look again at the original paper and give a careful analysis of its content.
This paper can be studied independently of [8]. Our notations are not quite the same, and we provide many more technical details. A discussion of Part I above is omitted, as this is a by now standard argument to assure that in order to obtain the desired result for arbitrary non-compact locally compact abelian groups, it suffices to consider only discrete ones.

The paper is organized as follows. In section 2 a description of the spectral synthesis problem and its history is given. Then in section 3 the above Part II is discussed. The most substantial part of the paper is section 4, where a detailed analysis is given of Malliavin's technical construction (Part III above). Finally in section 5 we make some complementary remarks concerning the differences in notation and in the proofs between this paper and [8]. We also give a list of corrections to [8]. Of course these are not meant to detract anything from the value of, or the admiration for Professor Malliavin's masterly work. They may however be useful if the reader wants to compare this paper with [8] - as he or she is invited to do.

We finally remark that a somewhat shorter version of part of this paper (roughly the sections 2 and 3) has appeared ([15]) as the text of a talk given at the Real Analysis Symposium in Coleraine, Northern Ireland, August 1988.

2 Description and notation

Let G be an infinite compact abelian group, and Γ its discrete dual (or character) group. Each $\gamma \in \Gamma$ determines a continuous character Θ_γ on G. On G we have the Haar measure dx, normalized so that $\int_G dx = 1$. On Γ we have the counting measure. Every $f \in L_1(G)$ has a Fourier transform $\hat{f} : \Gamma \to \mathbf{C}$, defined by

$$\hat{f}(\gamma) = \int_G f(x)\overline{\Theta_\gamma(x)}dx,$$

and hence a Fourier series,

$$\sum_{\gamma\in\Gamma} \hat{f}(\gamma)\Theta_\gamma.$$

Those f for which this series is absolutely convergent, i.e. for which $\hat{f} \in \ell_1(\Gamma)$, form the space $A(G)$. It is a uniformly dense subspace of $C(G)$, the space of all continuous functions on G. The dual space of $\ell_1(\Gamma)$ is $\ell_\infty(\Gamma)$. Its elements can be viewed as continuous linear functionals on $A(G)$. They are then called *pseudomeasures*, and their space is then denoted $PM(G)$. Each $\varphi \in \ell_\infty(\Gamma)$ determines a pseudomeasure $\hat{\varphi}$. The duality is given by

$$< f, \hat{\varphi} > = < \hat{f}, \varphi > = \sum_{\gamma\in\Gamma} \hat{f}(\gamma)\varphi(\gamma).$$

Finally, there is a natural way to define $\mathrm{Supp}(\hat{\varphi})$ as a closed subset of G: it is the complement of the largest open subset of G such that $< f, \hat{\varphi} >= 0$ for all $f \in A(G)$ with Supp (f) contained in that open set. Supp $(\hat{\varphi})$ is called the *spectrum* of φ.

Now the spectral synthesis problem can be formulated as follows. If E is a closed subset of G, if $g \in A(G)$ satisfies $g|_E = 0$ and if $\hat{\varphi} \in PM(G)$ satisfies $\mathrm{Supp}\hat{\varphi} \subset E$, does it then follow that $< g, \hat{\varphi} >= 0$? If so, then E is called a *set of spectral synthesis* or a *Wiener set* (cf. [9]). Certainly $< g, \hat{\varphi} >= 0$ is the case if $\hat{\varphi}$ is a measure; but not all pseudomeasures are measures. If $< g, \hat{\varphi} >\neq 0$ can occur, then spectral synthesis does not hold for Γ, and the associated set E is then called a *set of non spectral synthesis.*

We shall now give two other, equivalent, descriptions of the spectral synthesis problem. Although they will not be used in the rest of the paper, they may provide a wider perspective and a better understanding of the problem.
(a) The problem can be formulated entirely in terms of $\ell_\infty(\Gamma)$, as follows. Take $\varphi \in \ell_\infty(\Gamma)$; form the weak*-closed translation invariant subspace generated by φ; take all characters on Γ (i.e. elements of $\hat{G}$) contained in that space (actually, this is precisely the spectrum of φ, defined above). Now form the weak*-closed translation invariant subspace generated by these characters. Does that space contain φ? If so, then φ can be synthesized from its spectrum. The term "spectral synthesis", introduced by A. Beurling, is based on this formulation of the problem; cf. [9], p.140.

(b) The most usual way to formulate the problem is in terms of the Banach algebra $\ell_1(\Gamma)$, or, as we will do below, in terms of its Gel'fand representation, the Banach algebra $A(G)$, defined above. Each ideal in $A(G)$ has a *cospectrum*, namely the (closed) set of those points of G where all functions of the ideal vanish. Conversely, any closed subset E of G determines on the one hand the closed ideal I_E of all functions $f \in A(G)$ that vanish on E, and on the other hand the closed ideal J_E of all $f \in A(G)$ that can be approximated, in the norm of $A(G)$, by functions $g \in A(G)$ vanishing on a neighbourhood of E. If a function $\varphi \in \ell_\infty(\Gamma)$ has its spectrum contained in E, then, by a continuity argument, one has $< f, \hat{\varphi} >= 0$ for all $f \in J_E$. The spectral synthesis question can now be rephrased as: does $I_E = J_E$ hold?

Remark

If I_E and J_E are different, then there are automatically a great many ideals contained between them. In fact, if any two different closed ideals, one contained in the other, have the same cospectrum, then there is a third closed ideal enclosed between them (of course, again with the same cospectrum). This was proved in 1952 by H. Helson [4], and his result implies immediately the existence of at least countably many closed ideals between the two. That there are in fact continuum many closed ideals between them was proved by the author in 1966 [13]. Later that same year (cf. [9], p.36, footnote) it was also proved, for a more general class of function algebras, by Y. Katznelson (unpublished). In 1968 his proof appeared in [9] (p.36), and from there it has found its way into other books ([6], p.517, [1], pp.194-196).

We end this section with some additional remarks on the history of the subject. The beginning of spectral synthesis is usually placed in 1932, the year that N. Wiener proved his Tauberian theorem ([16]). Translated into the above terminology this theorem states, for the case $G = \mathbf{R}$, that every non-zero function $\varphi \in \ell_\infty(\Gamma)$ has a non-empty spectrum, or, equivalently, that the empty subset of $\mathbf{R}$ is a set of spectral synthesis.

A. Beurling was the first to state the spectral synthesis problem; not yet very explicitly in his paper [2] of 1938 (quoted in [8]), and more clearly in papers that appeared between 1945 and 1949. We refer to [6] for the references and for some comments on Beurling's papers (pp.601, 549-551).

The first set of non-synthesis was discovered in 1948 by L. Schwartz [12]: the sphere in $\mathbf{R}^3$. Apart from generalizations of Schwartz's result this remained the only example until 1959. In the opposite direction there were two important results by C.S. Herz: the ternary Cantor set on the line and the circle in the plane are of synthesis (1956 and 1958, respectively). We refer to the long survey article [5] by Herz on the subject, written largely before 1959.

For more information the reader is referred to the literature, e.g. [6], Chapter 10 (122 pages!), [9], [1] and [3], where also further references can be found.

3 The functional analytic argument

Let, for the time being, $f : G \to \mathbf{R}$ be a fixed real valued, but otherwise arbitrary, element of $A(G)$. Soon an extra condition (M) will be imposed, and the technical construction III is needed to show that it is possible to fulfill this condition.
Now we obtain the desired result in four steps.

3.1 Transfer of characters

Consider the real line $\mathbf{R}$, and its dual real line. The continuous characters on $\mathbf{R}$ are the functions $\chi_u(t) = e^{iut} (u \in \mathbf{R})$. By means of f these characters are transferred to G, by the definition: $F_u = \chi_u \circ f$, or, more explicitly: $F_u(x) = e^{iuf(x)}$. The $F_u (u \in \mathbf{R})$ resemble to a certain extent a group of characters on G. Indeed, they satisfy $|F_u| = 1$ and $F_u F_v = F_{u+v}$ (in particular, $F_0 = 1$ and $F_{-u} = \bar{F_u}$). So the F_u form a group homeomorphic to $\mathbf{R}$. However, the character relation $F_u(x)F_u(y) = F_u(x+y)$ does not hold. The Fourier coefficients of the F_u, thus the functions $\hat{F}_u$, will play a crucial rôle. One obviously has $\|\hat{F}_u\|_\infty \leq 1$, but the following much stronger smallness condition will be required:

$$(M) \qquad \int_{\mathbf{R}} \|\hat{F}_u\|_\infty |u| du < \infty.$$

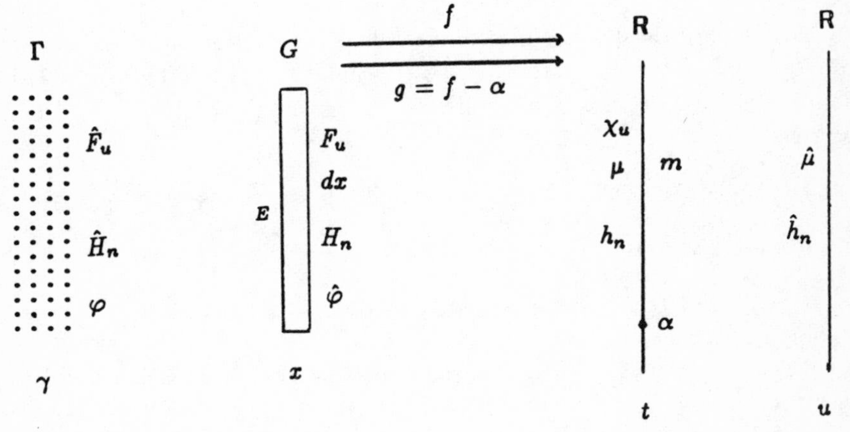

3.2 Transfer of Haar measure

The Haar measure dx on G is transferred by f to a measure μ on $\mathbf{R}$, in the following way: $\mu(k) = \int_{\mathbf{R}} k(t) d\mu(t) = \int_G k(f(x)) dx \qquad (k \in C(\mathbf{R}))$. We notice that $\mu \geq 0$, $\|\mu\| = 1$ and μ has compact support.
We can compute its Fourier-Stieltjes transform:

$$\hat{\mu}(-u) = \mu(\chi_u) = \int_G F_u(x) dx = \hat{F}_u(0).$$

Condition (M) (here not yet used in its full strength) implies that $\int_{\mathbf{R}} |u\hat{\mu}(u)|du < \infty$, and this, in turn, implies that, actually, $d\mu(t) = m(t)dt$ for some function $m \in C^1(\mathbf{R})$. The function m satisfies: $m \geq 0$, $\int_{\mathbf{R}} m(t)dt = 1$ and m has compact support.
We can now choose E and g (cf. section 2): take any $\alpha \in \mathbf{R}$ with $m(\alpha) > 0$ (there is at least an open interval of such α's). Take $E = f^{-1}(\{\alpha\})$ and take $g = f - \alpha$. Then $g|_E = 0$ and $g \in A(G)$ (in fact, $\hat{g}(\gamma) = \hat{f}(\gamma)$ for all $\gamma \neq 0$, and $\hat{g}(0) = \hat{f}(0) - \alpha$).
It remains to find a suitable function $\varphi \in \ell_\infty(\Gamma)$ with Supp $(\hat{\varphi}) \subset E$.

3.3 Transfer of derivators

Let $D : \mathbf{R} \to \mathbf{R}$ be the odd piecewise linear function defined by:

$$D(t) = D(2-t) = \frac{t}{2} \quad (0 \leq t \leq 1), \quad D(t) = 0 \quad (|t| \geq 2).$$

Define $h_n(t) = n^2 D(n(t-\alpha))$ (so that $\mathrm{Supp}(h_n) = [\alpha - \frac{2}{n}, \alpha + \frac{2}{n}]$ and Range $(h_n) = [-\frac{1}{2}n^2, \frac{1}{2}n^2]$).
It is a routine matter to see that for $k \in C^1(\mathbf{R})$ one has:

$$\lim_{n\to\infty} \int_{\mathbf{R}} k(t)h_n(t)dt = k'(\alpha).$$

Indeed, this follows easily by first checking the special case $k(t) = t - \alpha$:

$$\int_{\mathbf{R}} (t-\alpha)h_n(t)dt = \int_{-2}^{2} tD(t)dt = 2\int_0^1 (t + 2 - t)D(t)dt = 1.$$

Now we compute the Fourier transforms, where, to shorten the notation, we shall write $\frac{\sin u}{u} = \mathrm{SIN}\ u$. We start with the triangular function $T(t) = \max(1 - |t|, 0)$, whose Fourier transform is $\hat{T}(u) = \mathrm{SIN}^2 \frac{u}{2}$. We have $D(t) = \frac{1}{2}(T(t-1) - T(t+1))$, and hence

$$\hat{D}(u) = \frac{1}{2}(e^{-iu} - e^{iu})\mathrm{SIN}^2\frac{u}{2} = -i \sin u\ \mathrm{SIN}^2\frac{u}{2} = -iu \cos\frac{u}{2}\mathrm{SIN}^3\frac{u}{2}.$$

This, in turn, gives

$$\hat{h}_n(u) = n^2 \frac{1}{n}\hat{D}(\frac{u}{n})e^{-i\alpha u} = -iu\cos(\frac{u}{2n})\ \mathrm{SIN}^3(\frac{u}{2n})e^{-i\alpha u}.$$

As $\hat{h}_n(u) = O(|u|^{-2})$ $(|u| \to \infty)$, we can apply the Fourier inversion formula, which gives:

$$h_n(t) = \frac{1}{2\pi}\int_{\mathbf{R}} \hat{h}_n(u)\chi_u(t)du.$$

The functions h_n (the "derivators") are transferred by f into functions $H_n = h_n \circ f$, defined on G. The above formula for the h_n leads successively to:

$$H_n(x) = \frac{1}{2\pi}\int_{\mathbf{R}} \hat{h}_n(u)F_u(x)du,$$

$$\hat{H}_n(\gamma) = \frac{1}{2\pi}\int_{\mathbf{R}} \hat{h}_n(u)\hat{F}_u(\gamma)du.$$

We can now show that the $\hat{H}_n$ form a Cauchy sequence in $\ell_\infty(\Gamma)$. In fact, using the formula for $\hat{h}_n(u)$ we see that:

$$|\hat{H}_n(\gamma) - \hat{H}_m(\gamma)| \leq \frac{1}{2\pi} \int_{\mathbf{R}} |\hat{h}_n(u) - \hat{h}_m(u)| \|\hat{F}_u\|_\infty du \leq$$

$$\leq \frac{1}{2\pi} \int_{\mathbf{R}} 2|u| \|\hat{F}_u\|_\infty du.$$

So, if $\delta > 0$ is given, we can, due to condition (M), first take K so large that

$$\int_{|u|\geq K} |u| \|\hat{F}_u\|_\infty du < \delta,$$

and then take L so large that $n, m \geq L$ and $|u| \leq K$ implies:

$$|\hat{h}_n(u) - \hat{h}_m(u)| < \frac{\delta}{K}.$$

Then we will have $\|\hat{H}_n - \hat{H}_m\|_\infty < \delta$.
For φ we now take the uniform limit of the sequence $(\hat{H}_n)_{n\geq 1}$. Since

$$\mathrm{Supp}(H_n) \subset f^{-1}(\mathrm{Supp}(h_n)) = \{x \in G \mid |f(x) - \alpha| \leq \frac{2}{n}\},$$

it follows with a standard argument that

$$\mathrm{Supp}(\hat{\varphi}) \subset \{x \in G \mid f(x) = \alpha\} = E.$$

3.4 The proof

We have now all the ingredients needed to complete the proof. Indeed, we have

$$< \hat{g}, \hat{H}_n >= \int_G (f(x) - \alpha) H_n(x) dx = \int_{\mathbf{R}} (t - \alpha) h_n(t) m(t) dt.$$

When $n \to \infty$, the left hand term tends to $< \hat{g}, \varphi >$, whereas the right hand term converges to

$$\frac{d}{dt}((t - \alpha) m(t))|_{t=\alpha} = m(\alpha).$$

We conclude that $< \hat{g}, \varphi >= m(\alpha) \neq 0$, as desired.

4 The construction of f

We shall now describe the construction of a function $f : G \to \mathbf{R}$ of the form

$$f = \sum_{\gamma \in \Gamma} b_\gamma \Theta_\gamma \text{ with } \sum_{\gamma \in \Gamma} |b_\gamma| < \infty,$$

such that the corresponding functions $F_u = e^{iuf}$ satisfy condition (M). In fact, as in [8], a much stronger inequality (useful for the study of non-synthesis phenomena) will be obtained:

$$(\tilde{M}) \quad \|\hat{F}_u\|_\infty \leq e^{-a|u|^{\frac{1}{2}}} \quad \text{(some } a > 0, |u| \text{ sufficiently large).}$$

We first discuss how f will be made real-valued. If an element $\gamma \in \Gamma$ has order $r(\gamma) = 2$, then $\Theta_\gamma(x)$ takes only the values ± 1. Therefore b_γ will be taken real for such γ.
If, on the other hand, $r(\gamma) \geq 3$ (possibly $+\infty$), then γ and $-\gamma$ will be considered together and we will take $b_{-\gamma} = \overline{b_\gamma}$. Then

$$b_\gamma\Theta_\gamma + b_{-\gamma}\Theta_{-\gamma} = b_\gamma\Theta_\gamma + \overline{b_\gamma\Theta_\gamma} = 2Re(b_\gamma\Theta_\gamma).$$

We could for instance take $b_\gamma \in \mathbf{R}$ and obtain a summand $2b_\gamma Re\Theta_\gamma$. However, we follow [8] and decide to take b_γ purely imaginary. Then $2Re(b_\gamma\Theta_\gamma) = a_\gamma$ Im Θ_γ, where $a_\gamma = 2ib_\gamma$. If Γ possesses infinitely many elements of order 2, then we will use only such elements (i.e., we will take $b_\gamma = 0$ if $r(\gamma) \geq 3$). If not, then only elements of order at least 3 will be used (i.e., $b_\gamma = 0$ if $r(\gamma) = 2$).
In both cases ((A) and (B), say) we will now select a sequence $\gamma_1, \gamma_2, ...$ of suitable elements of Γ, and all b_γ, resp. a_γ, will be taken equal to zero, except the b_{γ_k}, resp. a_{γ_k}, which will be taken equal to $\frac{1}{k^2}$. Then the reality of f and the absolute summability of its Fourier series are clearly obtained, and the only remaining problem is to choose the γ_k in such a way that condition $(\tilde{M})$ is fulfilled.
We will henceforth abbreviate: $\Theta_{\gamma_k} = \Theta_k$. We will first discuss the relatively easy case (A). This case will also be helpful in understanding the more complicated cases that follow.

4.1 Case (A)

Choose any sequence $\gamma_1, \gamma_2, ...$ of independent elements of order 2 in Γ (i.e., $\sum_{j=1}^{k} n_j\gamma_j = 0$ with $n_j = 0$ or 1 implies $n_1 = ... = n_k = 0$).
We take $f = \sum_{k=1}^{\infty} \frac{1}{k^2}\Theta_k$.
For arbitrary $u \in \mathbf{R}$ and $x \in G$ we have

$$F_u(x) = e^{iuf(x)} = \prod_{k=1}^{\infty} \exp(i\frac{u}{k^2}\Theta_k(x)) =$$

$$= \prod_{k=1}^{\infty}(\cos\frac{u}{k^2} + i\Theta_k(x)\sin\frac{u}{k^2}) =$$

$$= \sum_{\Lambda}\left(\prod_{k\notin\Lambda}\cos\frac{u}{k^2}\right)\left(\prod_{k\in\Lambda} i\Theta_k(x)\sin\frac{u}{k^2}\right).$$

Here Λ ranges over all finite subsets of $\mathbf{N} = \{1, 2, 3, ...\}$ (including $\Lambda = \emptyset$). Indeed, when the infinite product is worked out, all terms with infinitely many sine-factors will vanish.

We put $\sum_{k\in\Lambda}\gamma_k = \gamma_\Lambda$ and we abbreviate $\Theta_{\gamma_\Lambda} = \Theta_\Lambda$; notice that $\Theta_\Lambda = \Pi_{k\in\Lambda}\Theta_k$. We then get:

$$F_u(x) = \sum_{\Lambda}\Theta_\Lambda(x)\left(\prod_{k\notin\Lambda}\cos\frac{u}{k^2}\right)\left(\prod_{k\in\Lambda} i\sin\frac{u}{k^2}\right).$$

If $\Lambda_1 \neq \Lambda_2$ then $\Theta_{\Lambda_1} \neq \Theta_{\Lambda_2}$ (because the γ_k are independent). We therefore know $\hat{F}_u$:

$$\hat{F}_u(\gamma_\Lambda) = (\prod_{k\notin\Lambda}\cos\frac{u}{k^2})(\prod_{k\in\Lambda} i\sin\frac{u}{k^2}),$$

$$\hat{F}_u(\gamma) = 0 \text{ for all other } \gamma \in \Gamma.$$

(In particular, $\hat{F}_u(0) = \Pi_{k=1}^{\infty} \cos \frac{u}{k^2}$). It follows that

$$\|\hat{F}_u\|_\infty = \sup_\Lambda \left(\prod_{k \notin \Lambda} |\cos \frac{u}{k^2}| \right) \left(\prod_{k \in \Lambda} |\sin \frac{u}{k^2}| \right).$$

All factors in this product are bounded by 1, and most are of the form $\cos v$ with small v. If $|v| \leq \frac{\pi}{4}$ then $|\sin v| \leq \cos v \leq e^{-\frac{1}{3}v^2}$. (Indeed, $e^{-\frac{1}{3}v^2} \geq 1 - \frac{1}{3}v^2 = 1 - \frac{1}{2}v^2 + \frac{1}{6}v^2 \geq 1 - \frac{1}{2}v^2 + \frac{1}{24}v^4 \geq \cos v$.) Now fix $u \in \mathbf{R}$. If $k \geq (\frac{4}{\pi}|u|)^{1/2} = A$, then $\frac{|u|}{k^2} \leq \frac{\pi}{4}$, and hence

$$\|\hat{F}_u\|_\infty \leq \prod_{k \geq A} \exp(-\frac{1}{3}\frac{u^2}{k^4}) = \exp(-\frac{1}{3}u^2 \sum_{k \geq A} \frac{1}{k^4}).$$

For $|u|$ sufficiently large (e.g. $|u| \geq 306$) we have

$$\sum_{k \geq A} \frac{1}{k^4} \geq \frac{1}{3(A+1)^3} \geq \frac{1}{5}|u|^{-3/2}.$$

Thus

$$\|\hat{F}_u\|_\infty \leq \exp(-\frac{1}{15}|u|^{1/2})$$

for $|u|$ sufficiently large. Thus f satisfies condition $(\tilde{M})$.

4.2 Case (B)

Our aim is to find suitable elements $\gamma_1, \gamma_2, \ldots$ (with $r(\gamma_k) \geq 3$), and then take $f = \sum_{k=1}^{\infty} \frac{1}{k^2} \operatorname{Im} \Theta_k$. We then have to consider:

$$F_u(x) = e^{iuf(x)} = \prod_{k=1}^{\infty} \exp(i\frac{u}{k^2} \operatorname{Im} \Theta_k(x)).$$

We therefore first investigate, for arbitrary $\gamma \in \Gamma$ with finite order (there is no need yet to exclude order 2 or 1), and arbitrary $v \in \mathbf{R}$, the function

$$\exp(iv \operatorname{Im} \Theta_\gamma).$$

Proposition 1.

If $\gamma \in \Gamma$ has order $r(\gamma) = r < \infty$, then

$$\exp(iv \operatorname{Im} \Theta_\gamma(x)) = \sum_{m=1}^{r} P_{m,r}(v)\Theta_\gamma^m(x) \quad (x \in G, v \in \mathbf{R}),$$

where the functions $P_{m,r} : \mathbf{R} \to \mathbf{C} (1 \leq m \leq r)$ are given by

$$P_{m,r}(v) = \frac{1}{r}\sum_{l=1}^{r} \exp(iv \sin \frac{2\pi l}{r}) \exp(-2\pi i \frac{ml}{r}).$$

Example.

If $r = 2$, then $P_{m,2}(v) = \frac{1}{2}((-1)^m + 1)(m = 1, 2)$, thus $P_{1,2}(v) = 0, P_{2,2}(v) = 1$. Further, $\operatorname{Im} \Theta_\gamma(x) = 0$ and $\Theta_\gamma^2(x) = 1$. So we find $1 = 0 + 1$.

Proof.

Fix $\gamma \in \Gamma$ with $r(\gamma) = r < \infty$. Consider the cyclic group $H = \{m\gamma \mid 1 \le m \le r\}$, and its dual group $\hat{H} = \{l\delta \mid 1 \le l \le r\}$, duality being given by $< m\gamma, l\delta >= \exp(2\pi i \frac{ml}{r})$. For fixed $v \in \mathbf{R}$ we consider the function $h : \hat{H} \to \mathbf{C}$, defined by

$$h(l\delta) = \exp(iv \operatorname{Im} \ < \gamma, l\delta >) = \exp(iv \sin \frac{2\pi l}{r}).$$

We compute its Fourier transform:

$$\hat{h}(m\gamma) = \frac{1}{r} \sum_{l=1}^{r} h(l\delta)\overline{< m\gamma, l\delta >} =$$

$$= \frac{1}{r} \sum_{l=1}^{r} \exp(iv \sin \frac{2\pi l}{r}) \exp(-2\pi i \frac{ml}{r}) = P_{m,r}(v).$$

The Fourier inversion formula yields:

$$h(l\delta) = \sum_{m=1}^{r} P_{m,r}(v) < m\gamma, l\delta > .$$

Now consider an arbitrary element $x \in G$. The mapping $m\gamma \mapsto \Theta_{m\gamma}(x)$ is a character on H. Therefore there is a specific l (depending on x) such that for all m we have $\Theta_\gamma^m(x) = \Theta_{m\gamma}(x) =< m\gamma, l\delta >$. It follows that

$$\exp(iv \operatorname{Im} \Theta_\gamma(x)) = h(l\delta) = \sum_{m=1}^{r} P_{m,r}(v) \Theta_\gamma^m(x).$$

This completes the proof.

We collect some properties of the $P_{m,r}(v)$:

Proposition 2.

(i) $|P_{m,r}(v)| \le 1$

(ii) If $m < r$ then $|P_{m,r}(v)| \le |v|$.

Proof.

(i) is trivial. As for (ii), we have $\exp(iv \sin \frac{2\pi l}{r}) = 1 + \epsilon_l$, with $|\epsilon_l| \le |v \sin \frac{2\pi l}{r}| \le |v|$. Using that $m \ne r$ we derive:

$$P_{m,r}(v) = \frac{1}{r} \sum_{l=1}^{r} (1 + \epsilon_l) \exp(-2\pi i \frac{ml}{r}) = \frac{1}{r} \sum_{l=1}^{r} \epsilon_l \exp(-2\pi i \frac{ml}{r}).$$

This gives: $|P_{m,r}(v)| \le \frac{1}{r} \sum_{l=1}^{r} |\epsilon_l| \le |v|$, as desired.

Proposition 3.

(i) $P_{r,r}(v) = \frac{1}{r} \sum_{l=1}^{r} \cos(v \sin \frac{2\pi l}{r})$

(ii) If $|v| \le \frac{\pi}{2}$ and $r \ge 3$, then $0 < P_{r,r}(v) \le \frac{1}{2}(1 + e^{-\frac{1}{12}v^2})$.

Proof.

(i) $P_{r,r}(v) = \frac{1}{r}\sum_{l=1}^{r} \exp(iv\sin\frac{2\pi l}{r}) = \frac{1}{r}\sum_{l=1}^{r} \cos(v\sin\frac{2\pi l}{r})$.

(ii) We use again (as in 4.1) the estimate: $\cos w \le e^{-\frac{1}{3}w^2}$, which is easily seen to hold even for all w with $|w| \le \frac{3}{2}\pi$. For $|v| \le \frac{\pi}{2}$ this gives

$$0 < P_{r,r}(v) \le \frac{1}{r}\sum_{l=1}^{r} \exp(-\frac{1}{3}v^2 \sin^2\frac{2\pi l}{r}).$$

For at least $\frac{1}{2}r$ of the r values of l we have $|\sin\frac{2\pi l}{r}| \ge \frac{1}{2}$. This is geometrically obvious for large r, and for small $r \ge 3$ it is easily verified (e.g., for $r = 5$ it holds for $l = 1,2,3,4$, because $\frac{2}{5} < \frac{5}{12}$). Note, however, that for $r = 2$ it is false. Anyway, for these values of l we have: $\exp(-\frac{1}{3}v^2\sin^2\frac{2\pi l}{r}) \le \exp(-\frac{1}{12}v^2)$, and for the other values of l 1 is a bound. Hence $0 < P_{r,r}(v) \le \frac{1}{r}(\frac{1}{2}r + \frac{1}{2}re^{-\frac{1}{12}v^2}) = \frac{1}{2}(1 + e^{-\frac{1}{12}v^2})$.

Remark.

In [8] (p. 65) it is stated that $P_{0,2}(v)$ (that is our $P_{2,2}(v)$) equals $\cos v$. But actually, as we have seen in the Example, $P_{2,2}(v) = 1$. Therefore the proof in [8] does not work for case (A), so this case has to be dealt with separately.

The functions $P_{m,r}(v)$ are closely related to the Bessel functions $J_n(z)(n \in \mathbf{Z}, z \in \mathbf{C})$. In fact we have:

Proposition 4.

$$P_{m,r}(v) = \sum_{k=-\infty}^{\infty} J_{m+kr}(v) \ (1 \le m \le r).$$

Proof.

We use the well known formula $e^{iz\sin t} = \sum_{n=-\infty}^{\infty} J_n(z)e^{int}$ to obtain:

$$P_{m,r}(v) = \frac{1}{r}\sum_{l=1}^{r}\Big(\sum_{n=-\infty}^{\infty} J_n(v)\exp(2\pi i\frac{nl}{r})\Big)\exp(-2\pi i\frac{ml}{r}).$$

Interchanging the order of summation we get:

$$P_{m,r}(v) = \sum_{n=-\infty}^{\infty} J_n(v)\frac{1}{r}\sum_{l=1}^{r}\exp(2\pi i(n-m)\frac{l}{r}).$$

This gives the result.

Lemma.

If $n \ge \frac{1}{4}v^2$ then $|J_n(v)| < (\frac{e|v|}{2n})^n$.

Proof.

We use the formula (valid for $n \geq 0$):

$$J_n(v) = (\frac{v}{2})^n \sum_{j=0}^{\infty} \frac{(-1)^j}{j!(n+j)!4^j} v^{2j} =$$

$$= (\frac{v}{2})^n \frac{1}{n!}(1 - \frac{v^2}{4(n+1)} + \frac{v^4}{32(n+1)(n+2)} - ...).$$

When $\frac{v^2}{4n} \leq 1$ this is an alternating series whose terms decrease in absolute value. Hence, for such v and n, $|J_n(v)| \leq (\frac{|v|}{2})^n \frac{1}{n!}$. Now $n! > (\frac{n}{e})^n$ (by Stirling's formula). The lemma follows.

Proposition 5.

If $|v| \geq 6$ then, for all $r \geq 1$:

$$\sum_{m=1}^{r} |P_{m,r}(v)| \leq v^2.$$

Proof.

From proposition 4 we obtain:

$$\sum_{m=1}^{r} |P_{m,r}(v)| \leq \sum_{k=-\infty}^{\infty} |J_k(v)| = |J_0(v)| + 2\sum_{k=1}^{\infty} |J_k(v)|.$$

Take the integer d such that $e|v| \leq d < e|v| + 1$. Then $\frac{e|v|}{2d} \leq \frac{1}{2}$, and hence, by the lemma:

$$2\sum_{k=d}^{\infty} |J_k(v)| \leq 2\sum_{k=d}^{\infty} 2^{-d} = 2^{2-d} \leq 1 \text{ (because } d \geq 2).$$

Using that $|J_k(v)| \leq 1$ for all $k \in \mathbf{Z}$, we get

$$\sum_{m=1}^{r} |P_{m,r}(v)| \leq 1 + 2(d-1) + 1 = 2d.$$

The proof is completed by observing that $|v| \geq 6$ implies: $2d < 2(e|v|+1) < 6|v| \leq v^2$.

In the next proposition it is shown that "the middle parts of the sums $\sum_{m=1}^{r} |P_{m,r}(v)|$ are small". As in [8] we let $|m| > d$ stand for $d < m < r - d$, the value of r being tacitly assumed.

Proposition 6.

For every $\eta > 0$ and $R > 0$ there is a positive integer d such that

$$\sum_{|m|>d} |P_{m,r}(v)| < \eta$$

for all v with $|v| \leq R$, and for all r.

Proof.

Just as in the previous proof we have, for any d:

$$\sum_{|m|>d} |P_{m,r}(v)| = \sum_{|m|>d} | \sum_{k=-\infty}^{\infty} J_{m+kr}(v)| \leq$$

$$\leq \sum_{|m|>d} \sum_{k=-\infty}^{\infty} |J_{m+kr}(v)| \leq 2 \sum_{n=d+1}^{\infty} |J_n(v)|.$$

Now take $d \geq \max(\frac{1}{4}R^2, eR, 1 - \frac{\log \eta}{\log 2})$. If $|v| \leq R$ then $d \geq \frac{1}{4}v^2$, thus, by the lemma:

$$\sum_{|m|>d} |P_{m,r}(v)| \leq 2 \sum_{n=d+1}^{\infty} (\frac{eR}{2d})^n \leq 2 \sum_{n=d+1}^{\infty} 2^{-n} = 2^{1-d} < \eta.$$

After all these preparations we shall describe how a suitable sequence of elements $\gamma_1, \gamma_2, ..$ in Γ can be chosen. Three cases will be distinguished:

(B1) Γ is a group of bounded order: $r(\Gamma) = r < \infty$, i.e., every element of Γ has order at most r;

(B2) Γ is a group of unbounded order, but each element of γ has finite order;

(B3) Γ possesses an element of infinite order (e.g. $\Gamma = \mathbf{Z}$).

We will treat these three cases separately.

4.3 Case (B1)

Definition.

A finite subset $\Delta = \{\gamma_1, \gamma_2, ..., \gamma_q\}$ (always tacitly implied: $|\Delta| = q$) of Γ is called *independent* if $\gamma_k \neq 0$ for all k and $\sum_{k=1}^{q} n_k \gamma_k = 0$ (with $n_k \in \mathbf{Z}$) implies $n_k \gamma_k = 0$ for all k.

Let $\Delta = \{\gamma_1, ..., \gamma_q\}$ be an independent set in Γ. For $p = 1, 2, ..$ we put

$$H_p = \{\gamma \in \Gamma \mid p\gamma = \sum_{k=1}^{q} n_k \gamma_k \neq 0 \text{ for suitable } n_k\}.$$

Then, clearly, $\cup_{p\geq 1} H_p$ is the set of all those elements in Γ that are dependent of Δ. Moreover, if Γ has finite order $r(\Gamma) = r$, then this set is $\cup_{1\leq p<r} H_p$. Now we have:

Proposition 7.

If Γ has bounded order, then there is an infinite set of independent elements in Γ.

Proof.

If not, then there would exist a finite set $\Delta = \{\gamma_1, \gamma_2, ..., \gamma_q\}$ in Γ such that $\cup_{1 \le p < r} H_p = \Gamma \setminus \{0\}$. H_1 is finite (in fact, $|H_1| \le r^q$). Now suppose that, for some $t < r, H_1, ..., H_{t-1}$ are finite, whereas H_t is infinite. Then there is at least one element $\sum_{k=1}^{q} n_k \gamma_k \ne 0$ such that $t\gamma = \sum_{k=1}^{q} n_k \gamma_k$ for infinitely many $\gamma \in \Gamma$. Say for $\delta_1, \delta_2, ...$ (all different). It follows that $t(\delta_1 - \delta_m) = 0$. Thus all $\delta_1 - \delta_m$ $(m \ge 2)$ have order $\le t$. Hence $\delta_1 - \delta_m \in \cup_{1 \le p < t} H_p$ (here we use the assumption that *all* non-zero elements are dependent of Δ). But the $H_p (p < t)$ are finite. This contradiction concludes the proof.

Remark.

An easy special case of this proposition was used already in the proof of Case (A).

We are now ready to settle case (B1). Choose an infinite sequence $\gamma_1, \gamma_2, ...$ in Γ such that for all $q \ge 1$, γ_{q+1} is independent of $\gamma_1, ..., \gamma_q$. We may suppose that $r(\gamma_k) = r_k \ge 3$ for all k (order 2 can occur only finitely often, by assumption; leave such elements out). As before we write $\Theta_{\gamma_k} = \Theta_k$. Take $f(x) = \sum_{k=1}^{\infty} \frac{1}{k^2} \operatorname{Im} \Theta_k(x)$. Then

$$F_u(x) = e^{iuf(x)} = \prod_{k=1}^{\infty} \exp(i\frac{u}{k^2} \operatorname{Im} \Theta_k(x)) =$$

$$\prod_{k=1}^{\infty} \sum_{m=1}^{r_k} P_{m,r_k}(\frac{u}{k^2}) \Theta_k^m(x) = \sum_{T \in \mathcal{T}} \prod_{k=1}^{\infty} P_{T(k),r_k}(\frac{u}{k^2}) \Theta_T(x).$$

Here $\mathcal{T} = \{T = (T(k))_{k \ge 1} \mid 1 \le T(k) \le r_k, T(k) \ne r_k \text{ only finitely often}\}$, $\gamma_T = \sum_{k=1}^{\infty} T(k)\gamma_k$, and $\Theta_T = \Theta_{\gamma_T} = \Pi_{k=1}^{\infty} \Theta_k^{T(k)}$, a well defined element of Γ, because it is in fact a finite product. Indeed, when the infinite product above is worked out, all products for which $T(k) \ne r_k$ occurs infinitely often will vanish, because of (ii) of Proposition 2. All $\Theta_T (T \in \mathcal{T})$ are different, as follows from the independency of the γ_k. Therefore we have:

$$\hat{F}_u(\gamma) = \prod_{k=1}^{\infty} P_{T(k),r_k}(\frac{u}{k^2}) \text{ if } \gamma = \gamma_T \text{ with } T \in \mathcal{T}$$

$$\hat{F}_u(\gamma) = 0 \text{ otherwise.}$$

It follows that

$$\|\hat{F}_u\|_\infty = \sup_T \prod_{k=1}^{\infty} |P_{T(k),k}(\frac{u}{k^2})|.$$

Take $\rho = \frac{1}{2}(1 + \exp(-\frac{1}{192})) = 0.9974.. < 1$. It follows from (ii) of Propositions 3 and 2 that $|P_{T(k),k}(\frac{u}{k^2})| \le \rho$ whenever $\frac{1}{4} \le \frac{|u|}{k^2} \le \frac{1}{2}$ ($\frac{1}{2}$ could in fact be enlarged to ρ). This happens when $\sqrt{2}|u|^{1/2} \le k \le 2|u|^{1/2}$. The above product contains therefore at least $\frac{1}{2}|u|^{1/2}$ factors $\le \rho$, provided that $|u|$ is sufficiently large, say $|u| \ge 136$. Thus $\|\hat{F}_u\|_\infty \le \rho^{\frac{1}{2}|u|^{1/2}} = e^{-a|u|^{1/2}}$ (if $|u|$ is sufficiently large), with $a = \frac{1}{2} \log \frac{1}{\rho} \simeq 0,0013$. Thus condition $(\tilde{M})$ is satisfied.

4.4 Case (B2)

It may now no longer be possible to find an independent sequence of elements in Γ. The elements $\gamma_1, \gamma_2, \ldots$ will be chosen successively, with fastly increasing orders $r_1, r_2, \ldots$. To do this, we first make the following choices, for each $q = 1, 2, 3, \ldots$:

(i) We define $R_q = e^q$.
We use this notation to make comparison with [8] easier, and to facilitate tracing the role of these constants in the construction.

(ii) We take positive real numbers $\eta_{q,k}(k \geq q)$ so small that

$$\prod_{k=q}^{\infty}(1 + \eta_{q,k}) < 1 + e^{-\sqrt{R_q}}.$$

(iii) We take positive integers $d_q(k)(k \geq q)$ so large that

$$\sum_{|m|>d_q(k)} |P_{m,r}(v)| < \eta_{q,k}$$

for all r and for all v with $|v| \leq R_q$. Such $d_q(k)$ exist by Proposition 6.

Now the sequence $\gamma_1, \gamma_2, \ldots$ is constructed as follows:
We take γ_1 such that $r_1 > 2d_1(1)$.
Next we take γ_2 such that $r_2 > 2d_1(2)r_1$ and $r_2 > 2d_2(2)$.
Generally, when $\gamma_1, \ldots, \gamma_{k-1}$ are chosen, then we take γ_k such that

$$r_k > \max_{1 \leq q \leq k}(2d_q(k)\Pi_{j=q}^{k-1} r_j).$$

We can now prove a substitute for independency:

Proposition 8.

With the above choices we have: if $\sum_{k=q}^{N} n_k\gamma_k = 0$ and $|n_k| \leq 2d_q(k)$ $(q \leq k \leq N)$, then $n_k = 0$ for all k.

Proof.

Suppose $\sum_{k=q}^{N} n_k\gamma_k = 0$ in a non-trivial way, i.e., with not all n_k equal to 0. We can assume that $n_N \neq 0$. Then $-n_N\gamma_N = \sum_{k=q}^{N-1} n_k\gamma_k$. Multiplying both sides with $\Pi_{k=q}^{N-1} r_k$ we get, because $r_k\gamma_k = 0$:

$$-(\prod_{k=q}^{N-1} r_k)n_N\gamma_N = 0.$$

Thus $(\Pi_{k=q}^{N-1} r_k)n_N$ is a multiple of r_N. However, $|n_N| \leq 2d_q(N)$, thus

$$(\prod_{k=q}^{N-1} r_k)|n_N| \leq 2d_q(N) \prod_{k=q}^{N-1} r_k < r_N.$$

As $n_N \neq 0$ this gives a contradiction.

We now start the usual computation. Define

$$f(x) = \sum_{k=1}^{\infty} \frac{1}{k^2} \operatorname{Im} \Theta_k(x),$$

where again $\Theta_k = \Theta_{\gamma_k}$. Then

$$F_u(x) = e^{iuf(x)} = \prod_{k=1}^{\infty} \exp(i\frac{u}{k^2} \operatorname{Im} \Theta_k(x)).$$

We now fix $u \in \mathbf{R}$. Let us write $\exp(i\frac{u}{k^2} \operatorname{Im} \Theta_k(x)) = E_k(x)$. Then $F_u = \prod_{k=1}^{\infty} E_k$. For an arbitrary integer $q \geq 1$ we define $F_{u,q} = \prod_{k=q}^{\infty} E_k$. We can then write:

$$F_u = \prod_{k=1}^{q-1} E_k . F_{u,q}.$$

Taking the Fourier transform we obtain:

$$\hat{F}_u = \hat{E}_1 * .... * \hat{E}_{q-1} * \hat{F}_{u,q},$$

and hence

$$(1) \quad \|\hat{F}_u\|_\infty \leq \prod_{k=1}^{q-1} \|\hat{E}_k\|_1 . \|\hat{F}_{u,q}\|_\infty .$$

We now make the following specific choice for q, assuming $|u| > 1$.

$$(2) \quad \log|u| \leq q < 1 + \log|u|.$$

We shall first estimate the factors $\|\hat{E}_k\|_1$. From Proposition 1 we know that

$$E_k(x) = \sum_{m=1}^{r_k} P_{m,r_k}(\frac{u}{k^2})\Theta_k^m(x).$$

This implies that

$$\|\hat{E}_k\|_1 = \sum_{m=1}^{r_k} |P_{m,r_k}(\frac{u}{k^2})|.$$

Using the second inequality in (2) we obtain:

$$\frac{|u|}{k^2} \geq \frac{|u|}{(q-1)^2} > \frac{|u|}{(\log|u|)^2} \geq 6$$

if $|u|$ is sufficiently large (e.g., $|u| \geq 152$). Hence, by Proposition 5, we have for these $u : \|\hat{E}_k\|_1 \leq \frac{u^2}{k^4} \leq u^2 \quad (1 \leq k < q)$.
We have thus obtained the following estimate:

$$(3) \quad \prod_{k=1}^{q-1} \|\hat{E}_k\|_1 \leq |u|^{2(q-1)} \leq e^{2\log^2|u|}.$$

To estimate the other factor in (1) we start just as in the case (B1).

$$F_{u,q}(x) = \prod_{k=q}^{\infty} \sum_{m=1}^{r_k} P_{m,r_k}(\frac{u}{k^2})\Theta_k^m(x) = \sum_{T\in\mathcal{T}} \{\prod_{k=q}^{\infty} P_{T(k),r_k}(\frac{u}{k^2})\}\Theta_T(x),$$

where, just as in (B1):

$$\mathcal{T} = \{T = (T(k))_{k\geq q} \mid 1 \leq T(k) \leq r_k, T(k) \neq r_k \text{ only finitely often}\},$$

and $\Theta_T = \prod_{k=q}^{\infty} \Theta_k^{T(k)}$.
However, the Θ_T may now no longer be all different. We therefore decompose $\mathcal{T}$ into disjoint classes $\mathcal{T}_\Lambda$, as follows.
Let $\mathcal{N}$ denote the family of all finite subsets of $\{q, q+1, ..\}$, including the empty set. For $\Lambda \in \mathcal{N}$ we define

$$\mathcal{T}_\Lambda = \{T \in \mathcal{T} \mid |T(k)| > d_q(k) \Leftrightarrow k \in \Lambda\}.$$

(As before, $|T(k)| > d$ means $d < T(k) < r_k - d$). Then, clearly, $\mathcal{T} = \cup_{\Lambda\in\mathcal{N}} \mathcal{T}_\Lambda$, and:

$$F_{u,q}(x) = \sum_{\Lambda\in\mathcal{N}} \sum_{T\in\mathcal{T}_\Lambda} (\prod_{k=q}^{\infty} P_{T(k)})\Theta_T(x),$$

where $P_{T(k)}$ is an abbreviation for $P_{T(k),r_k}(\frac{u}{k^2})$. We now split the infinite product, as follows. For $\Lambda \in \mathcal{N}$ we define

$$\mathcal{T}(\Lambda, >) = \{R = (R(k))_{k\in\Lambda} \mid |R(k)| > d_q(k)(\forall k \in \Lambda)\},$$

$$\mathcal{T}(\Lambda^c, \leq) = \{S = (S(k))_{k\notin\Lambda} \mid |S(k)| \leq d_q(k)(\forall k \notin \Lambda)\}.$$

It should perhaps be remarked that, conventionally, $\mathcal{T}(\emptyset, >)$ consists of one element: the empty sequence.
For a fixed $\Lambda \in \mathcal{N}$ we have then

$$\sum_{T\in\mathcal{T}_\Lambda} (\prod_{k=q}^{\infty} P_{T(k)})\Theta_T(x) =$$

$$= (\sum_{R\in\mathcal{T}(\Lambda,>)} (\prod_{k\in\Lambda} P_{R(k)})\Theta_R(x))(\sum_{S\in\mathcal{T}(\Lambda^c,\leq)} (\prod_{k\notin\Lambda} P_{S(k)})\Theta_S(x)),$$

where

$$\Theta_R = \prod_{k\in\Lambda} \Theta_k^{R(k)}, \Theta_S = \prod_{k\notin\Lambda} \Theta_k^{S(k)}.$$

Thus

$$F_{u,q} = \sum_{\Lambda\in\mathcal{N}} \sum_{R\in\mathcal{T}(\Lambda,>)} (\prod_{k\in\Lambda} P_{R(k)})\Theta_R\omega_\Lambda,$$

with

$$\omega_\Lambda = \sum_{S\in\mathcal{T}(\Lambda^c,\leq)} (\prod_{k\notin\Lambda} P_{S(k)})\Theta_S.$$

For a fixed $S \in \mathcal{T}(\Lambda^c, \leq)$ the characters Θ_S are all different. Indeed, suppose $\Theta_{S_1} = \Theta_{S_2}$ $(S_1, S_2 \in \mathcal{T}(\Lambda^c, \leq))$. Then we have, for N sufficiently large (taking $S_1(k) = S_2(k) = r_k$ if $k \in \Lambda$):

$$\sum_{k=q}^{N}(S_1(k) - S_2(k))\gamma_k = 0.$$

As $|S_i(k)| \leq d_q(k)$ $(i = 1, 2)$, we have $|S_1(k) - S_2(k)| \leq 2d_q(k)$ $(\mathrm{mod} r_k)$. Therefore Proposition 8 implies that $S_1(k) = S_2(k)$ for all $k \geq q$, thus $S_1 = S_2$.
It follows that

$$\hat{\omega}_\Lambda(\gamma) = \prod_{k \notin \Lambda} P_{S(k)} \text{ if } \gamma = \gamma_S \text{ with } S \in \mathcal{T}(\Lambda^c, \leq),$$

$$\hat{\omega}_\Lambda(\gamma) = 0 \text{ otherwise.}$$

(Of course, $\gamma_S = \sum_{k \notin \Lambda} S(k)\gamma_k$, and $\Theta_S = \Theta_{\gamma_S}$).
Thus

$$\|\hat{\omega}_\Lambda\|_\infty = \sup_{S \in \mathcal{T}(\Lambda^c, \leq)} \prod_{k \notin \Lambda} |P_{S(k)}|.$$

Using the trivial equality

$$\|(\Theta_R \omega_\Lambda)^\wedge\|_\infty = \|\hat{\omega}_\Lambda\|_\infty$$

we obtain:

$$\|\hat{F}_{u,q}\|_\infty \leq \sum_{\Lambda \in \mathcal{N}} \sum_{R \in \mathcal{T}(\Lambda, >)} \prod_{k \in \Lambda} |P_{R(k)}| \|\hat{\omega}_\Lambda\|_\infty.$$

For $\Lambda = \emptyset$ a more careful estimate of $\|\hat{\omega}_\Lambda\|_\infty$ will be needed than for the other Λ's. Therefore, adopting the notation in [8] (p.67), we denote $\omega_\emptyset$ by ψ (omitting the subscript q) and $F_{u,q} - \psi$ by ω. Then we have

$$\|\hat{F}_{u,q}\|_\infty \leq \|\hat{\psi}\|_\infty + \|\hat{\omega}\|_\infty.$$

Here

$$\|\hat{\psi}\|_\infty = \sup_S \prod_{k \geq q} |P_{S(k), r_k}(\frac{u}{k^2})|,$$

the supremum being over all sequences $S = (S(k))_{k \geq q}$ that satisfy $|S(k)| \leq d_q(k)$ $(\forall k \geq q)$. Now, as in (B1), at least $\frac{1}{2}|u|^{1/2}$ factors in this product are smaller than $\rho < 1$. Thus

$$\|\hat{\psi}\|_\infty \leq \rho^{1/2|u|^{1/2}} = e^{-a|u|^{1/2}},$$

with $a = \frac{1}{2}\log\frac{1}{\rho}$, as before.
It remains to estimate $\|\hat{\omega}\|_\infty$. Here the trivial estimate $\|\hat{\omega}_\Lambda\|_\infty \leq 1$ will suffice. We get, using the properties (iii) and (ii) of the $d_q(k)$ and $\eta_{q,k}$:

$$\|\hat{\omega}\|_\infty \leq \sum_{\emptyset \neq \Lambda \in \mathcal{N}} \sum_{R \in \mathcal{T}(\Lambda, >)} \prod_{k \in \Lambda} |P_{R(k)}| =$$

$$= \sum_{\emptyset \neq \Lambda \in \mathcal{N}} \prod_{k \in \Lambda} \sum_{|m| > d_q(k)} |P_{m, r_k}(\frac{u}{k^2})| \leq$$

$$\leq \sum_{\emptyset \neq \Lambda \in \mathcal{N}} \prod_{k \in \Lambda} \eta_{q,k} = \prod_{k \geq q}(1 + \eta_{q,k}) - 1 <$$

$$< e^{-\sqrt{R_q}}.$$

Remark.

In [8] it is claimed (p.67, middle) that ω itself already satisfies $\|\omega\|_\infty < \exp(-R_q^{1/2})$ (whence, a fortiori, $\|\hat{\omega}\|_\infty$ also satisfies this inequality). To draw this conclusion, however, we would need, in the preceding estimate, that $\|\omega_\Lambda\|_\infty \le 1$ for every non-empty $\Lambda \in \mathcal{N}$, thus that:

$$\sum_{S \subset \mathcal{T}(\Lambda^c, \le)} \prod_{k \notin \Lambda} |P_{S(k), r_k}(\frac{u}{k^2})| \le 1.$$

Actually, however, one can only assert that each summand in this expression is bounded by 1. One therefore has to go through the more careful analysis above to obtain the desired estimate for $\|\hat{\omega}\|_\infty$. This inaccuracy in [8] was pointed out already in [14] (loose leaflet, Stelling VIII).

From the first inequality in (2) it follows that $|u| \le e^q = R_q$ and hence we can conclude:

$$\|\hat{\omega}\|_\infty \le e^{-|u|^{1/2}} < e^{-a|u|^{1/2}},$$

and thus

$$\|\hat{F}_{u,q}\|_\infty \le 2e^{-a|u|^{1/2}}.$$

Using (1) and (3) this gives

$$\|\hat{F}_u\|_\infty \le 2e^{\log^2 |u|} e^{-a|u|^{1/2}} \le e^{-b|u|^{1/2}}$$

for a fixed $b > 0$ and $|u|$ sufficiently large (e.g. $b = 0.001$ and $|u| \ge 10^{13}$).

4.5 Case (B3)

This case can be treated similarly. We only give some indications. Let $\gamma \in \Gamma$ have order $r(\gamma) = \infty$. Then we have, for arbitrary $v \in \mathbf{R}$:

$$\exp(iv \operatorname{Im} \Theta_\gamma(x)) = \sum_{m=-\infty}^{\infty} P_m(v) \Theta_\gamma^m(x),$$

where

$$P_m(v) = J_m(v) = \frac{1}{2\pi} \int_0^{2\pi} e^{iv \sin\varphi} e^{-im\varphi} d\varphi.$$

The following properties are used:

(1) $|P_m(v)| \le 1 \quad (m \in \mathbf{Z}, v \in \mathbf{R})$

(2) $|P_m(v)| \le |v| \quad (m \ne 0, v \in \mathbf{R})$

(3) $0 < P_0(v) < \frac{1}{2}(1 + e^{-\frac{1}{12}v^2}) \quad (|v| \le \frac{1}{2})$

(4) $|P_m(v)| \le \rho \quad (\text{some } \rho < 1; \frac{1}{4} \le |v| \le \frac{1}{2})$

(5) For all $\eta > 0$ and $R > 0$ there is an integer d such that $\sum_{|m|>d} |P_m(v)| < \eta$ for all $v \in \mathbf{R}$ with $|v| \le R$.

(6) $\sum_{m=-\infty}^{\infty} |P_m(v)| \leq v^2$ ($|v|$ sufficiently large)

We take numbers $R_q, \eta_{q,k}, d_q(k)$ as in case (B2). Now take $\gamma_k = \lambda_k \gamma (k \geq 1)$, where the λ_k are positive integers such that

$$\lambda_k > 2 \sum_{j=q}^{k-1} \lambda_j d_q(j) \quad (1 \leq q \leq k).$$

If $\sum_{k=q}^{N} n_k \lambda_k \gamma = 0$ and $|n_k| \leq 2d_q(k)$ for all $q \leq k \leq N$, then $n_k = 0$ for all these k. For indeed, $\sum_{k=q}^{N} n_k \lambda_k \gamma = 0$ is possible only if $\sum_{k=q}^{N} n_k \lambda_k = 0$. But then we have:

$$|n_N \lambda_N| = |\sum_{k=q}^{N-1} n_k \lambda_k| \leq \sum_{k=q}^{N-1} |n_k| \lambda_k \leq \sum_{k=q}^{N-1} 2d_q(k) \lambda_k < \lambda_N.$$

Thus $n_N = 0$. Likewise all the n_k equal 0.
The remaining part of the proof of $(\tilde{M})$ now follows exactly as in case (B2).

5 Complementary remarks

5.1 Notation

To make comparison easier we list some differences of notation. We write:

$$\Gamma,\ G,\ \gamma,\ x,\ \Theta_\gamma(x),\ f,\ F_u,$$

where in [8] the following symbols are used:

$$G,\ G',\ g,\ g',\ < g, g' >,\ \varphi,\ e^{iu\varphi}.$$

In [8] a distinction in notation is made for the Fourier transformation on $\mathbf{R}(h \mapsto \hat{h})$ and that on $G(h \mapsto h')$, whereas we use $\hat{h}$ in both cases. Further, for $k \in L^1(G)$ our $\|\hat{k}\|_\infty$ is denoted $\|k\|'_\infty$ in [8], and likewise with the subscript 1.

5.2 Misprints

For the reader's convenience we include a list of misprints in [8]:
p.63, l.11 f.b.: $|\hat{h}_n(u)|$ instead of $|h_n(u)|$; p.64, l.2: $\bar{b}$ instead of b; p.65, l.2: v instead of m, v; p.66, l.3 f.b.: "r et v" instead of "m et r et v"; p.67, l.5: "est divisé par" instead of "divise"; l.7: $< mg_k, g' >$ instead of Im $< g_k, g' >$; l.12: Im $< g_k, g' >$ instead of $< g_k, g' >$; l.14: $\|\omega(g')\|'_\infty$ instead of $\|\omega(g')\|_\infty$; l.16: $|P_{m,s}(v)|$ instead of $|P_{m,r}(v)|$.

5.3 Inaccuracies

There are some further inaccuracies in [8]. On p.65, l.18 "$P_{0,2}(v) = \cos v$" should be "$P_{0,2}(v) = 1$"(cf. our Remark in 4.2). This, however, invalidates the rest of the sentence. For that reason we had to consider case (A) separately.
The inequality $\|\omega(g')\|_\infty < \exp(-R_q^{1/2})$ (p.67, l.13) is not correct (cf. our Remark in 4.4). As we have shown in 4.4, the following "a fortiori" conclusion in itself is correct (after the dash is added, as indicated in 5.2).

5.4 Further remarks

(i) In section 3.3 we make a specific choice for the functions h_n, and we compute the $\hat{h}_n$ explicitly. In this way we don't need to make an appeal to the theory of distributions. Moreover it enables us to show that the $\hat{H}_n$ are not merely weak*, but uniformly convergent in $\ell_\infty(\Gamma)$. Cf. p.63 of [8].

(ii) Our proof of the lemma in 4.2 is different from that suggested on p.64 of [8].

(iii) On p.66 of [8] a diagonal procedure is used to select the elements γ_k. This, however, can easily be avoided, as we have done in 4.4. We have further made there an explicit choice for the constants R_q straightaway.

References

[1] J.J. Benedetto - Spectral synthesis. Stuttgart: B.G. Teubner 1975.

[2] A. Beurling - Sur les intégrales de Fourier absolument convergentes et leur application à une transformation fonctionnelle. IXe Congrès des Math. Scand., Helsinki, August 1938, 345-366 (1938).

[3] C.C. Graham, O.C. McGehee - Essays in commutative harmonic analysis. New York, etc.: Springer-Verlag 1979.

[4] H. Helson - On the ideal structure of group algebras. Ark.Mat. 2, 83-86 (1952).

[5] C.S. Herz - The spectral theory of bounded functions. Transactions Amer. Math. Soc. 94, 181-232 (1960).

[6] E. Hewitt, K.A. Ross - Abstract harmonic analysis, Vol. II. Berlin etc.: Springer-Verlag 1970.

[7] J.-P. Kahane, R. Salem - Ensembles parfaits et séries trigonométriques. Paris: Hermann 1963.

[8] P. Malliavin - Impossibilité de la synthèse spectrale sur les groupes abéliens non compacts. Inst. Hautes Etudes Sci. Publ. Math. 2, 61-68 (1959).

[9] H. Reiter - Classical harmonic analysis and locally compact groups. Oxford: At the Clarendon Press 1968.

[10] I. Richards - On the disproof of spectral synthesis. J. Comb. Theory 2, 61-70 (1967).

[11] W. Rudin - Fourier analysis on groups. Interscience Tract No. 12. New York: Wiley 1962.

[12] L. Schwartz - Sur une propriété de synthèse spectrale dans les groupes non compacts. C.R. Acad. Sci. Paris 227, 424-426 (1948).

[13] J.D. Stegeman - Extension of a theorem of H. Helson. Int. Congress Math., Moscow, Abstracts, Section 5, p.28 (1966).

[14] J.D. Stegeman - Studies in Fourier and tensor algebras. Thesis. Utrecht: Pressa Trajectina 1971.

[15] J.D. Stegeman - On Malliavin's proof of non spectral synthesis. Real Analysis Exchange 14 (1), 30-33 (1988 - 1989).

[16] N. Wiener - Tauberian theorems. Annals of Math. 33, 1-100 (1932).

Department of Mathematics
University of Utrecht
P.O. Box 80010
3508 TA Utrecht
The Netherlands

Papers of G. Cross, Y. Kubota, J.L. Mawhin, M. Morayne, W.F. Pfeffer and W.-C. Yang, and C.A. Rogers

Papers delivered at the conference by G. Cross (U. Waterloo), Y. Kubota (Ibaraki U.), J.L. Mawhin (U. Louvain-la-neuve), M. Morayne (U. Wrocław), W.-C. Yang (U. California, Davis), and C.A. Rogers (University College London) have not been included in this volume. The references are:

1. J. Cichoń, M. Morayne, J. Pawlikowski, ***Decomposing Baire functions***, Real Analysis Exchange 14 (1988–89) 16–19.

2. G. Cross, ***Generalized integrals as limits of Riemann-like sums***, Real Analysis Exchange 13 (1987–88) 390–403.

3. Y. Kubota, ***Extension of the Denjoy integral***, not yet published.

4. J.L. Mawhin, ***Applications of multiple generalised Riemann integrals***, not yet published.

5. W.F. Pfeffer and W.-C. Yang, ***A multidimensional integral and its extensions***, to appear in Real Analysis Exchange.

6. C.A. Rogers, ***Dimension prints***, Mathematica 35 (1988) 1–27.

Problems

1. (**Lee P.Y., University of Singapore**) Let the space of Henstock integrable functions on $[a, b]$ be called the Denjoy space D. It is known that D can be normed by

$$\|f\| = \sup\{|\int_a^x f(t)dt| : a \leq x \leq b\}.$$

Given a set X in the Denjoy space D with the given norm, find conditions such that X is compact, or compact in some other sense. In other words, characterize compact sets in D.

2. (**Lee P.Y., University of Singapore**) We conjecture that if f_n for $n = 1, 2, 3, \ldots$, and f are all Henstock integrable on $[a, b]$, $f_n(x) \to f(x)$ pointwise almost everywhere as $n \to \infty$ and

$$\int_a^b f_n(x)g(x)dx \to \int_a^b f(x)g(x)dx \text{ as } n \to \infty$$

whenever g is of bounded variation on $[a, b]$, then the primitives F_n of f_n are ACG^* uniformly in n.

3. (**V.A. Skvortsov, Moscow State University**) Prove (or disprove) that the Perron integral, defined for an abstract derivative base by means of its major and minor functions continuous with respect to this base, is equivalent to the integral for the same base but defined without the precondition of continuity of major and minor functions. If the answer is negative, describe the class of bases for which this equivalence holds true.

4. (**S. James Taylor, University of Virginia**) Does there exist a non-countable set $A \subset \mathbb{R}$ such that $\forall B \subset \mathbb{R}$ with Lebesgue measure zero, $\Lambda(A \times B) = 0$? (Λ denotes Hausdorff linear measure.) Is this true for Pfeffer 'small' subsets or 'slight' subsets? [It is known that, for each B with $|B| = 0$, there is a perfect set A, depending on B, such that $\Lambda(A \times B) = 0$.]

LECTURE NOTES IN MATHEMATICS
Edited by A. Dold and B. Eckmann

Some general remarks on the publication of proceedings of congresses and symposia

Lecture Notes aim to report new developments - quickly, informally and at a high level. The following describes criteria and procedures which apply to proceedings volumes. The editors of a volume are strongly advised to inform contributors about these points at an early stage.

§1. One (or more) expert participant(s) of the meeting should act as the responsible editor(s) of the proceedings. They select the papers which are suitable (cf. §§ 2, 3) for inclusion in the proceedings, and have them individually refereed (as for a journal). It should not be assumed that the published proceedings must reflect conference events faithfully and in their entirety. Contributions to the meeting which are not included in the proceedings can be listed by title. The series editors will normally not interfere with the editing of a particular proceedings volume - except in fairly obvious cases, or on technical matters, such as described in §§ 2, 3. The names of the responsible editors appear on the title page of the volume.

§2. The proceedings should be reasonably homogeneous (concerned with a limited area). For instance, the proceedings of a congress on "Analysis" or "Mathematics in Wonderland" would normally not be sufficiently homogeneous.

One or two longer survey articles on recent developments in the field are often very useful additions to such proceedings - even if they do not correspond to actual lectures at the congress. An extensive introduction on the subject of the congress would be desirable.

§3. The contributions should be of a high mathematical standard and of current interest. Research articles should present new material and not duplicate other papers already published or due to be published. They should contain sufficient information and motivation and they should present proofs, or at least outlines of such, in sufficient detail to enable an expert to complete them. Thus resumes and mere announcements of papers appearing elsewhere cannot be included, although more detailed versions of a contribution may well be published in other places later.

Surveys, if included, should cover a sufficiently broad topic, and should in general not simply review the author's own recent research. In the case of surveys, exceptionally, proofs of results may not be necessary.

"Mathematical Reviews" and "Zentralblatt für Mathematik" require that papers in proceedings volumes carry an explicit statement that they are in final form and that no similar paper has been or is being submitted elsewhere, if these papers are to be considered for a review. Normally, papers that satisfy the criteria of the Lecture Notes in Mathematics series also satisfy this

.../...

requirement, but we would strongly recommend that the contributing authors be asked to give this guarantee explicitly at the beginning or end of their paper. There will occasionally be cases where this does not apply but where, for special reasons, the paper is still acceptable for LNM.

§4. Proceedings should appear soon after the meeeting. The publisher should, therefore, receive the complete manuscript within nine months of the date of the meeting at the latest.

§5. Plans or proposals for proceedings volumes should be sent to one of the editors of the series or to Springer-Verlag Heidelberg. They should give sufficient information on the conference or symposium, and on the proposed proceedings. In particular, they should contain a list of the expected contributions with their prospective length. Abstracts or early versions (drafts) of some of the contributions are very helpful.

§6. Lecture Notes are printed by photo-offset from camera-ready typed copy provided by the editors. For this purpose Springer-Verlag provides editors with technical instructions for the preparation of manuscripts and these should be distributed to all contributing authors. Springer-Verlag can also, on request, supply stationery on which the prescribed typing area is outlined. Some homogeneity in the presentation of the contributions is desirable.

Careful preparation of manuscripts will help keep production time short and ensure a satisfactory appearance of the finished book. The actual production of a Lecture Notes volume normally takes 6 -8 weeks.

Manuscripts should be at least 100 pages long. The final version should include a table of contents and as far as applicable a subject index.

§7. Editors receive a total of 50 free copies of their volume for distribution to the contributing authors, but no royalties. (Unfortunately, no reprints of individual contributions can be supplied.) They are entitled to purchase further copies of their book for their personal use at a discount of 33.3 %, other Springer mathematics books at a discount of 20 % directly from Springer-Verlag. Contributing authors may purchase the volume in which their article appears at a discount of 33.3 %.

Commitment to publish is made by letter of intent rather than by signing a formal contract. Springer-Verlag secures the copyright for each volume.

Vol. 1259: F. Cano Torres, Desingularization Strategies for Three-Dimensional Vector Fields. IX, 189 pages. 1987.

Vol. 1260: N.H. Pavel, Nonlinear Evolution Operators and Semigroups. VI, 285 pages. 1987.

Vol. 1261: H. Abels, Finite Presentability of S-Arithmetic Groups. Compact Presentability of Solvable Groups. VI, 178 pages. 1987.

Vol. 1262: E. Hlawka (Hrsg.), Zahlentheoretische Analysis II. Seminar, 1984–86. V, 158 Seiten. 1987.

Vol. 1263: V.L. Hansen (Ed.), Differential Geometry. Proceedings, 1985. XI, 288 pages. 1987.

Vol. 1264: Wu Wen-tsün, Rational Homotopy Type. VIII, 219 pages. 1987.

Vol. 1265: W. Van Assche, Asymptotics for Orthogonal Polynomials. VI, 201 pages. 1987.

Vol. 1266: F. Ghione, C. Peskine, E. Sernesi (Eds.), Space Curves. Proceedings, 1985. VI, 272 pages. 1987.

Vol. 1267: J. Lindenstrauss, V.D. Milman (Eds.), Geometrical Aspects of Functional Analysis. Seminar. VII, 212 pages. 1987.

Vol. 1268: S.G. Krantz (Ed.), Complex Analysis. Seminar, 1986. VII, 195 pages. 1987.

Vol. 1269: M. Shiota, Nash Manifolds. VI, 223 pages. 1987.

Vol. 1270: C. Carasso, P.-A. Raviart, D. Serre (Eds.), Nonlinear Hyperbolic Problems. Proceedings, 1986. XV, 341 pages. 1987.

Vol. 1271: A.M. Cohen, W.H. Hesselink, W.L.J. van der Kallen, J.R. Strooker (Eds.), Algebraic Groups Utrecht 1986. Proceedings. XII, 284 pages. 1987.

Vol. 1272: M.S. Livšic, L.L. Waksman, Commuting Nonselfadjoint Operators in Hilbert Space. III, 115 pages. 1987.

Vol. 1273: G.-M. Greuel, G. Trautmann (Eds.), Singularities, Representation of Algebras, and Vector Bundles. Proceedings, 1985. XIV, 383 pages. 1987.

Vol. 1274: N. C. Phillips, Equivariant K-Theory and Freeness of Group Actions on C*-Algebras. VIII, 371 pages. 1987.

Vol. 1275: C.A. Berenstein (Ed.), Complex Analysis I. Proceedings, 1985–86. XV, 331 pages. 1987.

Vol. 1276: C.A. Berenstein (Ed.), Complex Analysis II. Proceedings, 1985–86. IX, 320 pages. 1987.

Vol. 1277: C.A. Berenstein (Ed.), Complex Analysis III. Proceedings, 1985–86. X, 350 pages. 1987.

Vol. 1278: S.S. Koh (Ed.), Invariant Theory. Proceedings, 1985. V, 102 pages. 1987.

Vol. 1279: D. Ieşan, Saint-Venant's Problem. VIII, 162 Seiten. 1987.

Vol. 1280: E. Neher, Jordan Triple Systems by the Grid Approach. XII, 193 pages. 1987.

Vol. 1281: O.H. Kegel, F. Menegazzo, G. Zacher (Eds.), Group Theory. Proceedings, 1986. VII, 179 pages. 1987.

Vol. 1282: D.E. Handelman, Positive Polynomials, Convex Integral Polytopes, and a Random Walk Problem. XI, 136 pages. 1987.

Vol. 1283: S. Mardešić, J. Segal (Eds.), Geometric Topology and Shape Theory. Proceedings, 1986. V, 261 pages. 1987.

Vol. 1284: B.H. Matzat, Konstruktive Galoistheorie. X, 286 pages. 1987.

Vol. 1285: I.W. Knowles, Y. Saitō (Eds.), Differential Equations and Mathematical Physics. Proceedings, 1986. XVI, 499 pages. 1987.

Vol. 1286: H.R. Miller, D.C. Ravenel (Eds.), Algebraic Topology. Proceedings, 1986. VII, 341 pages. 1987.

Vol. 1287: E.B. Saff (Ed.), Approximation Theory, Tampa. Proceedings, 1985–1986. V, 228 pages. 1987.

Vol. 1288: Yu. L. Rodin, Generalized Analytic Functions on Riemann Surfaces. V, 128 pages, 1987.

Vol. 1289: Yu. I. Manin (Ed.), K-Theory, Arithmetic and Geometry. Seminar, 1984–1986. V, 399 pages. 1987.

Vol. 1290: G. Wüstholz (Ed.), Diophantine Approximation and Transcendence Theory. Seminar, 1985. V, 243 pages. 1987.

Vol. 1291: C. Mœglin, M.-F. Vignéras, J.-L. Waldspurger, Correspondances de Howe sur un Corps p-adique. VII, 163 pages. 1987

Vol. 1292: J.T. Baldwin (Ed.), Classification Theory. Proceedings, 1985. VI, 500 pages. 1987.

Vol. 1293: W. Ebeling, The Monodromy Groups of Isolated Singularities of Complete Intersections. XIV, 153 pages. 1987.

Vol. 1294: M. Queffélec, Substitution Dynamical Systems – Spectral Analysis. XIII, 240 pages. 1987.

Vol. 1295: P. Lelong, P. Dolbeault, H. Skoda (Réd.), Séminaire d'Analyse P. Lelong – P. Dolbeault – H. Skoda. Seminar, 1985/1986. VII, 283 pages. 1987.

Vol. 1296: M.-P. Malliavin (Ed.), Séminaire d'Algèbre Paul Dubreil et Marie-Paule Malliavin. Proceedings, 1986. IV, 324 pages. 1987.

Vol. 1297: Zhu Y.-l., Guo B.-y. (Eds.), Numerical Methods for Partial Differential Equations. Proceedings. XI, 244 pages. 1987.

Vol. 1298: J. Aguadé, R. Kane (Eds.), Algebraic Topology, Barcelona 1986. Proceedings. X, 255 pages. 1987.

Vol. 1299: S. Watanabe, Yu.V. Prokhorov (Eds.), Probability Theory and Mathematical Statistics. Proceedings, 1986. VIII, 589 pages. 1988.

Vol. 1300: G.B. Seligman, Constructions of Lie Algebras and their Modules. VI, 190 pages. 1988.

Vol. 1301: N. Schappacher, Periods of Hecke Characters. XV, 160 pages. 1988.

Vol. 1302: M. Cwikel, J. Peetre, Y. Sagher, H. Wallin (Eds.), Function Spaces and Applications. Proceedings, 1986. VI, 445 pages. 1988.

Vol. 1303: L. Accardi, W. von Waldenfels (Eds.), Quantum Probability and Applications III. Proceedings, 1987. VI, 373 pages. 1988.

Vol. 1304: F.Q. Gouvêa, Arithmetic of *p*-adic Modular Forms. VIII, 121 pages. 1988.

Vol. 1305: D.S. Lubinsky, E.B. Saff, Strong Asymptotics for Extremal Polynomials Associated with Weights on $\mathbb{R}$. VII, 153 pages. 1988.

Vol. 1306: S.S. Chern (Ed.), Partial Differential Equations. Proceedings, 1986. VI, 294 pages. 1988.

Vol. 1307: T. Murai, A Real Variable Method for the Cauchy Transform, and Analytic Capacity. VIII, 133 pages. 1988.

Vol. 1308: P. Imkeller, Two-Parameter Martingales and Their Quadratic Variation. IV, 177 pages. 1988.

Vol. 1309: B. Fiedler, Global Bifurcation of Periodic Solutions with Symmetry. VIII, 144 pages. 1988.

Vol. 1310: O.A. Laudal, G. Pfister, Local Moduli and Singularities. V, 117 pages. 1988.

Vol. 1311: A. Holme, R. Speiser (Eds.), Algebraic Geometry, Sundance 1986. Proceedings. VI, 320 pages. 1988.

Vol. 1312: N.A. Shirokov, Analytic Functions Smooth up to the Boundary. III, 213 pages. 1988.

Vol. 1313: F. Colonius, Optimal Periodic Control. VI, 177 pages. 1988.

Vol. 1314: A. Futaki, Kähler-Einstein Metrics and Integral Invariants. IV, 140 pages. 1988.

Vol. 1315: R.A. McCoy, I. Ntantu, Topological Properties of Spaces of Continuous Functions. IV, 124 pages. 1988.

Vol. 1316: H. Korezlioglu, A.S. Ustunel (Eds.), Stochastic Analysis and Related Topics. Proceedings, 1986. V, 371 pages. 1988.

Vol. 1317: J. Lindenstrauss, V.D. Milman (Eds.), Geometric Aspects of Functional Analysis. Seminar, 1986–87. VII, 289 pages. 1988.

Vol. 1318: Y. Felix (Ed.), Algebraic Topology – Rational Homotopy. Proceedings, 1986. VIII, 245 pages. 1988

Vol. 1319: M. Vuorinen, Conformal Geometry and Quasiregular Mappings. XIX, 209 pages. 1988.

Vol. 1380: H.P. Schlickewei, E. Wirsing (Eds.), Number Theory, Ulm 1987. Proceedings. V, 266 pages. 1989.

Vol. 1381: J.-O. Strömberg, A. Torchinsky. Weighted Hardy Spaces. V, 193 pages. 1989.

Vol. 1382: H. Reiter, Metaplectic Groups and Segal Algebras. XI, 128 pages. 1989.

Vol. 1383: D.V. Chudnovsky, G.V. Chudnovsky, H. Cohn, M.B. Nathanson (Eds.), Number Theory, New York 1985–88. Seminar. V, 256 pages. 1989.

Vol. 1384: J. Garcia-Cuerva (Ed.), Harmonic Analysis and Partial Differential Equations. Proceedings, 1987. VII, 213 pages. 1989.

Vol. 1385: A.M. Anile, Y. Choquet-Bruhat (Eds.), Relativistic Fluid Dynamics. Seminar, 1987. V, 308 pages. 1989.

Vol. 1386: A. Bellen, C.W. Gear, E. Russo (Eds.), Numerical Methods for Ordinary Differential Equations. Proceedings, 1987. VII, 136 pages. 1989.

Vol. 1387: M. Petković, Iterative Methods for Simultaneous Inclusion of Polynomial Zeros. X, 263 pages. 1989.

Vol. 1388: J. Shinoda, T.A. Slaman, T. Tugué (Eds.), Mathematical Logic and Applications. Proceedings, 1987. V, 223 pages. 1989.

Vol. 1000: Second Edition. H. Hopf, Differential Geometry in the Large. VII, 184 pages. 1989.

Vol. 1389: E. Ballico, C. Ciliberto (Eds.), Algebraic Curves and Projective Geometry. Proceedings, 1988. V, 288 pages. 1989.

Vol. 1390: G. Da Prato, L. Tubaro (Eds.), Stochastic Partial Differential Equations and Applications II. Proceedings, 1988. VI, 258 pages. 1989.

Vol. 1391: S. Cambanis, A. Weron (Eds.), Probability Theory on Vector Spaces IV. Proceedings, 1987. VIII, 424 pages. 1989.

Vol. 1392: R. Silhol, Real Algebraic Surfaces. X, 215 pages. 1989.

Vol. 1393: N. Bouleau, D. Feyel, F. Hirsch, G. Mokobodzki (Eds.), Séminaire de Théorie du Potentiel Paris, No. 9. Proceedings. VI, 265 pages. 1989.

Vol. 1394: T.L. Gill, W.W. Zachary (Eds.), Nonlinear Semigroups, Partial Differential Equations and Attractors. Proceedings, 1987. IX, 233 pages. 1989.

Vol. 1395: K. Alladi (Ed.), Number Theory, Madras 1987. Proceedings. VII, 234 pages. 1989.

Vol. 1396: L. Accardi, W. von Waldenfels (Eds.), Quantum Probability and Applications IV. Proceedings, 1987. VI, 355 pages. 1989.

Vol. 1397: P.R. Turner (Ed.), Numerical Analysis and Parallel Processing. Seminar, 1987. VI, 264 pages. 1989.

Vol. 1398: A.C. Kim, B.H. Neumann (Eds.), Groups – Korea 1988. Proceedings. V, 189 pages. 1989.

Vol. 1399: W.-P. Barth, H. Lange (Eds.), Arithmetic of Complex Manifolds. Proceedings, 1988. V, 171 pages. 1989.

Vol. 1400: U. Jannsen. Mixed Motives and Algebraic K-Theory. XIII, 246 pages. 1990.

Vol. 1401: J. Steprāns, S. Watson (Eds.), Set Theory and its Applications. Proceedings, 1987. V, 227 pages. 1989.

Vol. 1402: C. Carasso, P. Charrier, B. Hanouzet, J.-L. Joly (Eds.), Nonlinear Hyperbolic Problems. Proceedings, 1988. V, 249 pages. 1989.

Vol. 1403: B. Simeone (Ed.), Combinatorial Optimization. Seminar, 1986. V, 314 pages. 1989.

Vol. 1404: M.-P. Malliavin (Ed.), Séminaire d'Algèbre Paul Dubreil et Marie-Paul Malliavin. Proceedings, 1987 – 1988. IV, 410 pages. 1989.

Vol. 1405: S. Dolecki (Ed.), Optimization. Proceedings, 1988. V, 223 pages. 1989.

Vol. 1406: L. Jacobsen (Ed.), Analytic Theory of Continued Fractions III. Proceedings, 1988. VI, 142 pages. 1989.

Vol. 1407: W. Pohlers, Proof Theory. VI, 213 pages. 1989.

Vol. 1408: W. Lück, Transformation Groups and Algebraic K-Theory. XII, 443 pages. 1989.

Vol. 1409: E. Hairer, Ch. Lubich, M. Roche. The Numerical Solution of Differential-Algebraic Systems by Runge-Kutta Methods. VII, 139 pages. 1989.

Vol. 1410: F.J. Carreras, O. Gil-Medrano, A.M. Naveira (Eds.), Differential Geometry. Proceedings, 1988. V, 308 pages. 1989.

Vol. 1411: B. Jiang (Ed.), Topological Fixed Point Theory and Applications. Proceedings, 1988. VI, 203 pages. 1989.

Vol. 1412: V.V. Kalashnikov, V.M. Zolotarev (Eds.), Stability Problems for Stochastic Models. Proceedings, 1987. X, 380 pages. 1989.

Vol. 1413: S. Wright, Uniqueness of the Injective III$_1$ Factor. III, 108 pages. 1989.

Vol. 1414: E. Ramírez de Arellano (Ed.), Algebraic Geometry and Complex Analysis. Proceedings, 1987. VI, 180 pages. 1989.

Vol. 1415: M. Langevin, M. Waldschmidt (Eds.), Cinquante Ans de Polynômes. Fifty Years of Polynomials. Proceedings, 1988. IX, 235 pages. 1990.

Vol. 1416: C. Albert (Ed.), Géométrie Symplectique et Mécanique. V, 289 pages. 1990.

Vol. 1417: A.J. Sommese, A. Biancofiore, E.L. Livorni (Eds.). Algebraic Geometry. Proceedings, 1988. V, 320 pages. 1990.

Vol. 1418: M. Mimura, Homotopy Theory and Related Topics. Proceedings, 1988. V, 241 pages. 1990.

Vol. 1419: P.S. Bullen, P.Y. Lee, J.L. Mawhin, P. Muldowney, W.F. Pfeffer (Eds.), New Integrals. Proceedings, 1988. V, 202 pages. 1990.